A DECADE OF *Barbie*™ DOLLS AND COLLECTIBLES

1981 – 1991

IDENTIFICATION & VALUES

BETH SUMMERS

COLLECTOR BOOKS

A Division of Schroeder Publishing Co., Inc.

Searching for a Publisher?

We are always looking for knowledgeable people considered to be experts within their fields. If you feel that there is a real need for a book on your collectible subject and have a large comprehensive collection, contact Collector Books.

Cover design: Beth Summers
Book design: Michelle Dowling
Photography: Charley Lynch

CONTENTS

DEDICATION

This book is dedicated to my husband, B.J. Over the course of the past years he has spent many days touring the Barbie product aisles of toy stores. He has stood in those pink corridors, just him, and maybe two or three little 8 or 9 year old girls, searching for a particular item from my "want" list. Finding the doll, he has turned to the little girls and looking into their sweet, curious, and somewhat puzzled little faces said, "I've looked everywhere for this doll." Then, walking away, he has heard the sounds of giggles and some guesses about why a grown man would buy a Barbie doll!

He has assisted me in transporting collectibles to and from the photography studio and helped me in so many ways that took up a great deal of time throughout this project. Being a collector in other fields of interest, he understands the compulsion to amass great quantities of whatever it is you find captivating. He is my best friend and a truly special person. I am incredibly lucky to be his wife.

ACKNOWLEDGMENTS

This project would not have come about if it were not for three very special friends. Deborah Summers and Robbie Spees rekindled my interest in the Barbie doll and introduced me to Roszella Jones. That is when we formed our club, the Western Kentucky Barbie Doll Collectors. Friend Deana Jones joined when I saw her in the Barbie products aisle at a department store and we began talking about our common interest in collecting. Deborah, Robbie, and Deana are great friends and have made my collecting hobby so much fun!

Photographing dolls still in their boxes presents a unique challenge requiring someone with a love of photography and the knowledge to do the job well. Avoiding the reflection of light off the clear plastic windows in the packaging of these collectibles was a concern of mine when this project began. Charley Lynch has done an excellent job photographing these collectibles and with his help, we have compiled a wonderful array of Barbie doll memorabilia for you to view in a beautiful full-color format.

An incredible amount of time was spent poring over piles and piles of value lists for these collectibles from all over the United States. Especially helpful to me were:

Marl & B

The collector's premier source for Barbie dolls, fashions, and accessories from 1959 to the present. Marl produces a quarterly catalog of collectibles. If you wish to buy, sell or trade collectibles or want a catalog of articles for sale, you may write or call:

Marl Davidson

10301 Branden Run ▸ Bradenton, Florida 34202

Phone: 941-751-6275 ▸ Fax: 941-751-5463

Kitty's Collectibles

Featuring Barbie dolls, she is the publisher of a bi-monthly 100+ page quarter million dollar sales list and produces fantastic all-Barbie Doll Shows and Sales. To buy or sell Barbie collectibles, or to receive show or catalog information you may write or call:

Kitty's Collectibles

6568 Modoc Lane ▸ Evergreen, Colorado 80439

Phone: 303-670-9312 ▸ Fax: 303-670-9276

AOL Address: KittysCol@AOL.COM ▸ Web page: HTTP://Users.AOL.COM/KittysCol/Kittys.HTM

SPECIAL RECOGNITION ≡

The most crucial step in this project was finding the actual items. It is through the gracious cooperation of Roszella Jones, "The Barbie Doll Lady of Paducah," that this was accomplished. Roszella is the founder and vice president of the Western Kentucky Barbie Doll Collectors, a club interested not only in Barbie doll, but also active in contributing to the surrounding community. She is an avid collector and active dealer. She has had numerous newspaper articles written about her and her special interest, and is featured in the November/December 1995 issue of *Barbie Bazaar*.

Tours of Girl Scouts, school groups, garden clubs, church groups, homemakers clubs, community groups, and just anyone interested in Barbie dolls have passed through her home to view room after room of her treasured collectibles. She is continually altering her collection by selling or trading to obtain new or different items. She has been gathering Barbie dolls and memorabilia for over thirty-two years and has a truly awesome collection. Counted among her favorites in this decade are the 1989 Happy Holidays® Barbie® doll and Spiegel's Sterling Wishes Barbie. Roszella allowed me to transport her collectibles to a studio for photography and then return them. It was an exhaustive process and I wish to thank Roszella for enduring this task so cheerfully. I also want to thank her husband, Ewell, for packing dolls in and out of his home and remaining such a good sport about this whole project.

If you wish to contact Roszella to buy, sell, trade or just talk Barbie, you may write to her at:

Roszella Jones
c/o Collector Books ▸ P.O. Box 3009 ▸ Paducah, KY 42003-3009

Surrounding Roszella in this photograph are centerpieces from two of the conventions sponsored and produced by the Western Kentucky Barbie Doll Collectors. These special collectibles were designed and created by Deborah Summers. The two centerpieces at the left are from the first convention, "Barbie in the Land of Enchantment." (If you wish to know more about this convention, see the club news article in the September/October 1994 issue of *Barbie Bazaar*.) The centerpiece to the right is from the most recent convention in the fall of 1995, "Barbie Tours the Orient." (If you wish to know more about this convention, see the club news article in the July/August 1996 issue of *Barbie Bazaar*.)

Roszella Jones, "The Barbie Doll Lady of Paducah."

INTRODUCTION

By now almost everyone knows that Ruth Handler's idea for an adult figure doll came from observing her daughter playing with fashion paper dolls. For hours on end her daughter would be entertained changing fashions on these cardboard models. I can remember as a child spending wonderful hours with movie star paper dolls like Patty Page, Doris Day and more. The clothing that was printed for these paper doll packages were fashions current to the time period. I was so excited about the glamorous fashions that I, too, would play tirelessly with these wonderful paper dolls. So Ruth's idea that little girls were interested in contemporary fashions proved to be the beginning of an absolute phenomenon in the toy industry.

1981 to 1991 was a bounteous decade for Barbie dolls and collectibles. These years saw the introduction of the Happy Holidays Series and continuation of the International Series. These two series have gone on to unimaginable success. In 1984 Andy Warhol painted Barbie in his most famous style, introducing this doll to the prestigious art world. Bob Mackie and Oscar de la Renta also entered the Barbie doll fashion scene. The Summit Barbie doll commemorated the first annual Barbie Summit in 1990 with children from many countries participating in a unique cultural exchange. Audrey Hepburn played a part in the production of the UNICEF Barbie doll, with 37¢ of the sale price of each going to the U.N. committee for this organization. Porcelain dolls required a higher investment and took Barbie dolls to a new collecting level. At the end of this decade, the Barbie logo was revised.

It is interesting to note that not many years ago, Barbie dolls were not even mentioned in doll collecting circles. For some reason this toy icon of the early sixties was not considered worthy of a collector's time or attention. Some of the early books on collecting Barbie doll items were so unpopular that they were just sitting on warehouse shelves and editors were trying to figure out ways to give them away.

Then suddenly the market began to change. It may have been the popular International Series or the Happy Holiday Series that brought attention back to Barbie doll. The celebration of the 25th year of Barbie doll production also stirred interest in the doll world. Or it may have been that many baby boomers who grew up with Barbie doll began to reminisce about the wonderful childhood times spent with this doll. Prices on antique dolls escalated so far out of the reach of many collectors that a new market began to form. Items produced in limited quantities just for collectors were becoming very popular.

Whatever the reason for the resurgence of interest in the Barbie doll, Mattel's primary market and the collectible secondary market are tightly tied together and Barbie doll's popularity has reached an all-time high. This doll has stirred controversy and been criticized as being a poor role model for little girls. But I know of no other doll that has portrayed women in the roles of doctor, astronaut, veterinarian, business executive, television reporter, teacher, firefighter, artist, actress, model, and more. Ethnic races of all kinds have been represented and Mattel has promoted the idea that "we girls can do anything!"

This book will be an on-going project for each decade of Barbie doll production back to the beginning in 1959. With this book and the continuing volumes, 1970 to 1980 and 1959 to 1969, I want to present as many photos of dolls, fashions, furniture, family, friends, animals, and accessories as I can find. A picture is worth a thousand words and you can see details in these photographs that could possibly be overlooked in written descriptions. With a virtual photo album of collectibles, you may be able to identify some of the things you have but didn't know exactly the origin.

Some dolls and articles covered in this first volume may have been produced a little before or a little after the decade of 1981–1991. Because items that have a copyright of a given year were not usually released until the following year, I have included the copyright year 1980. Also added is the copyright year of 1991 even though the doll or item may not have been released until 1992.

If you have unusual items relating to the Barbie doll collecting field, I would be happy to hear from you. I am now compiling information for the next volume, 1970 to 1980, and solicit information, photographs, or the use of your collectible to photograph here. You may contact me through the Collector Books office or at my home address: 233 Darnell Road, Benton, KY 42025. If you wish a reply, please enclose a SASE.

As I continue to gather information, I hope to help collectors both experienced and novice identify and evaluate their treasures.

PRICING

It is certainly a challenge to put together a book in any collectible field. I wanted this to be a report to you, the collector, on dolls that are on the secondary market and to supply as fair a market value as is possible to obtain. Pricing in this field is so volatile that arriving at a value is not an easy task. Because I am not a dealer, I am in an unbiased position to report values from across the country. I have taken figures from the East Coast, West Coast, Southern states and the Heartland and averaged them to come up with the values here. If I could find market prices for a given article that were consistent from several sources, I reported that value. If the amount varied quite a bit, then I averaged all of them for a single given item. For instance, if a West Coast value for a doll was $50.00, a Southern value $30.00, and a Heartland value $100.00, I added the three, then divided that total ($180.00) by three. The resulting value will be $60.00. So the values reported in this book are neither the highest nor the lowest found.

There have been reports of articles being sold at outrageous prices. Sometimes these amounts have come from auction results. You must remember that an amount given at an auction very often does not reflect the actual value of the collectible, but rather the fact that there are two collectors who do not want to give up on it, no matter what the cost! Some prices seen at shows or secondary market booths reflect an inflated price because the dealer did not want to part with the item, but also did not want to list it as not for sale. So, it was shown at a much higher price and if it did sell, it would be worth the time and effort to replace. There are also instances of some values actually having gone down. It may well be that these items were overpriced from the beginning.

Whether you agree or not, a doll removed from the box is worth much less, sometimes half. There are always exceptions to every rule. A good example of unboxed items retaining their values are the porcelain dolls or Mackie dolls. These and some others do not lose their values when removed from their boxes. Condition of the item determines the price as well. If items have been exposed to extreme heat, cold, sunlight, or cigarette smoke, they will deteriorate. Store them carefully if you plan to sell them at a later date. Many of these dolls were meant for little girls' playtime and have been a part of hours of fun-time fantasy. If pieces are missing or worn, the purchase price will go down drastically.

All the values given in this guide are for complete, mint, or never removed from box articles. If you have only parts of a collectible, you must divide and reduce your buying or selling value accordingly. If the items are worn from handling or otherwise impaired from exposure to sunlight, cigarette smoke or other environmental hazards, the value will be markedly reduced. If a value for a doll is

$100.00 NRFB (never removed from box), but the doll you have is without the box and missing a few pieces, it may be worth only $10.00. What a difference! This is a point that is very important to understand and yet seems to be the most misunderstood.

Of course, if the doll was yours or your little girl's, the memories connected with it are priceless! You can make yourself crazy boxing up and storing dolls to protect their monetary value. I feel very strongly that this hobby can be a charming pleasure to be shared with fellow collectors. Unless you are accumulating items for investment, I urge you to collect only what you really like and display them where you can view and take delight in them every day. Any doll or collectible can be sold at more or less the reported value, depending on the circumstances of the sale, but hopefully, this guide will give you a starting point. Please remember that the prices reported here are usually averages of price ranges found.

Condition Abbreviations

NRFB: Never Removed From Box. This term is self explanatory, but to command the corresponding price, the item and the box itself must be in excellent shape.

NRFP: Never Removed From Packaging. Some items came in a plastic package, plastic-coated cardboard or blister pack. To command the corresponding price, the item and the packaging itself must be in excellent shape.

Mint: This item may be removed from its original box, but it is in excellent shape, has been stored or displayed properly and has all its original accessories. This term will also apply to items such as books or comics that were never packaged, but sent in bulk to retail outlets. To command the corresponding price, the item must be in excellent shape, not played with or in any way deteriorated.

DOLLS

The most popular item for collecting is the doll itself. The Barbie doll evolved only slightly over the years 1981 to 1991, with alterations in the pose of the arms and hands being used interchangeably throughout the time span. It is important to note here that even if you have a doll with the copyright date of 1966 on the back of the torso, that does not mean it was produced in 1966. Even the Totally Hair Barbie doll with a box copyright date of 1991 has a torso with the 1966 copyright. That is the same torso mold that has been used throughout this time period and back to 1966 and continues to be used today. If you are in an antique shop and an out-of-box doll is tagged as a 1966 doll, be careful not to be misled. You may be looking at a brand new doll.

This third decade held a lot of excitement for Mattel. Bob Mackie entered the doll field with his Golden Mackie Barbie doll which became an instant hit. The porcelain line of dolls began with the exquisite Blue Rhapsody Barbie doll. International Dolls continued to gain popularity and became a mainstay throughout this decade. A wonderful line of Barbie and the Rockers dolls was introduced and the fifties theme was used for the Sensations line of dolls. Sears, Spiegel, F.A.O. Schwarz, Hills, Toys "Я" Us, K-Mart, Wal-Mart, Service Merchandise, Target and more department, toy, and specialty stores sponsored limited edition dolls. From high fashion designer and porcelain dolls to the My First Barbie dolls, the primary market retail price for 1981 to 1991 had quite a range.

The reported values on the secondary market for this section and all of this book are taken from across the country and averaged. The resulting value will be neither the highest nor the lowest found. If the doll is removed from its original packaging, the price will drop by half. If it is removed and was played with or has parts missing, the value will continue to go down. As with all of the fields, but especially this one, prices are for never removed from box (NRFB) items (see pricing explanation pages 10-11). To command the averaged values reported here, the box as well as the doll must be like new.

Rozella Jones Collection

Author's Collection

Rozella Jones Collection

Clockwise from top left:

ALL AMERICAN BARBIE #9423
This All American doll is dressed in red, white, and blue with a pink cotton knit crop top and comes with 2 pairs of Reebok Hi-tops.
©1990   NRFB $19.00

ALL AMERICAN KIRA #9427
This All American Kira doll is dressed in a denim jacket accented with gold stars, bright yellow crop top, red, white, and blue vest, and denim shorts, and comes with 2 pairs of Reebok Hi-tops.
©1990   NRFB $19.00

BARBIE AND THE ALL STARS BARBIE #9099
This doll is dressed in a pink body suit with a bodice, briefs, and leggings of white material spangled with blue metallic stars. This package also includes a gym bag, skirt, workout weights, and more.
©1989   NRFB $17.00

BARBIE AND THE ALL STARS MIDGE #9360
Midge doll wears a purple and white baseball uniform and comes with a baseball bat and ball. Also included in this package is a tote bag that converts to a short dress for Midge.
©1989   NRFB $19.00

Author's Collection

Rozella Jones Collection

Rozella Jones Collection

Clockwise from top left:

ANGEL FACE BARBIE #5640
This doll wears a very Victorian-style outfit with a white lace top with high neck and puffed sleeves, black belt and satiny pink skirt. Her hair is unusual since it is sunstreaked.
©1982   NRFB $35.00

ANIMAL LOVIN' BARBIE/BLACK #4824
Barbie doll is dressed in a unique pink, black, and metallic gold fashion. Also in this package the doll is holding a loveable black and white panda.
©1988   NRFB $20.00

ANIMAL LOVIN' BARBIE/WHITE #1350
Barbie doll is dressed in a unique pink, black, and metallic gold fashion. Also in this package the doll is holding a loveable black and white panda.
©1988   NRFB $23.00

ARMY BARBIE #3966
This is the American Beauties Collection Limited Edition Army Barbie doll. She is dressed in an authentic Army officer's evening uniform.
©1989   NRFB $38.00

Roszella Jones Collection

Clockwise from top left:

ASTRONAUT BARBIE/BLACK   **#1207**
This astronaut outfit is pink and silver with a clear bubble helmet.
Barbie doll has her own flag to plant on the moon. Computer and
space maps are just a few of the extras included in this package.
©1985   NRFB $75.00

ASTRONAUT BARBIE/WHITE   **#2449**
This astronaut outfit is pink and silver with a clear bubble helmet.
Barbie doll has her own flag to plant on the moon. Computer and
space maps are just a few of the extras included in this package.
©1985   NRFB $80.00

BALLERINA BARBIE   **#4983**
This doll is dressed in a white ballerina costume with a white and silver tutu and comes with white toe shoes. She has a silver crown and
specially posed arms and hands.
©1983   NRFB $60.00

BALLROOM BEAUTY BARBIE   **#3678**
This is a Wal-Mart special limited edition. Barbie doll is dressed in a
lavender and iridescent ballgown that can be changed to six different
dance looks.
©1991   NRFB $38.00

Roszella Jones Collection

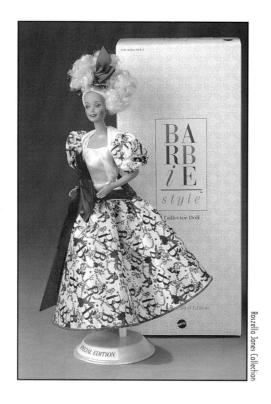

Rozella Jones Collection

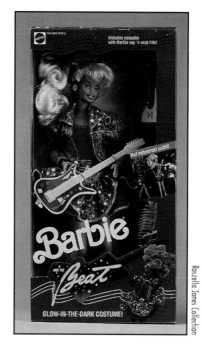

Rozella Jones Collection

Rozella Jones Collection

Rozella Jones Collection

Clockwise from top left:

BARBIE STYLE   #5315
This doll was first in the short-lived Applause series. She is wearing a party dress with a full skirt made from a bright pink, purple, blue, and yellow printed satin. A large pink bow is placed at her waist and her hair is pulled up by a similar bow which holds a tumble of blonde curls on the top of her head.

©1990   NRFB $38.00

BARBIE AND THE BEAT   #2751
This doll is dressed in a pink and blue denim jacket and ruffled skirt with glow-in-the-dark trim. She comes with a guitar and cassette with Barbie Rap 'n Rock hits.

©1989   NRFB $30.00

BATHTIME FUN BARBIE   #9601
This Barbie doll is dressed in a teal, pink, yellow, and purple maillot that has a blue foam skirt and bracelet. A bottle of foam soap in a choice of pink, blue, or yellow is included in this package.

©1990   NRFP $14.00

BEACH BLAST BARBIE   #3237
This doll comes in a hot pink two-piece bathing suit with sunglasses and Frisbee disc. She also has a hairpiece that changes colors in the sun from blonde to pink.

©1989   NRFB $18.00

Roszella Jones Collection

Roszella Jones Collection

Roszella Jones Collection

Clockwise from top left:

BLOSSOM BEAUTY BARBIE #3142
This is a Special Edition doll for Shop-Ko/Venture and is dressed in a bright purple, magenta, white, and teal gown accented with necklace, ring and earrings.
1991   NRFB $49.00

BLUE RHAPSODY BARBIE #1364
Not to be confused with the porcelain doll of the same name, this Special Edition doll for Service Merchandise is wearing a gown with a skirt made of layers of deep blue tulle with glittering gold accents and a gold bodice. Her hair bow is from the same blue and gold material.
©1991   NRFB $275.00

CALIFORNIA DREAM BARBIE #4439
This doll comes with a bathing suit, skirt, shirt, visor, a comic book, and a special Beach Boys record written just for Barbie doll.
©1987   Mint $20.00

Rozella Jones Collection

Author's Collection

Rozella Jones Collection

Clockwise from top left:

CELEBRATION BARBIE #2998
This special Barbie doll celebrates Sears' 100th Anniversary and was a Limited Edition. She wears a pink and silver ball gown and has a silver jumpsuit, too! A doll stand is also included in this package.
©1985 NRFB $86.00

BENEFIT BALL BARBIE #1521
First in the Classique collection, this doll models a gorgeous fashion of teal and gold. Extras include teal stockings and a necklace to accent the gown. Her titian hair is the perfect choice for this exquisite doll.
©1991 NRFB $116.00

COOL LOOKS BARBIE #5947
Dressed in hot pink, orange, and black, this Toys "Я" Us doll includes extras like a clock, journal, telephone, slumber case, pencil, soda can, and more.
1990 NRFB $25.00

Clockwise from top right:

COOL TIMES BARBIE   #3022

This doll comes with an ice cream soda and a scooter. She is wearing a bright pink satin jacket, short skirt, and white leggings with black polka dots.

©1988   NRFB $20.00

COSTUME BALL BARBIE   #7123

This doll is dressed in a pink and iridescent fashion that will convert to three styles. A tiny mask for the doll and a larger mask for a little girl are included in this package.

©1990   NRFB $26.00

CRYSTAL BARBIE/BLACK   #4859

This doll is dressed in a white and iridescent fashion with a full ruffled boa. Her jewelry is a pair of crystal drop earrings and matching necklace. A child-size crystal style Barbie medallion is included in the package.

©1983   NRFB $22.00

Roszella Jones Collection

Roszella Jones Collection

Roszella Jones Collection

Rozella Jones Collection

Rozella Jones Collection

Rozella Jones Collection

Rozella Jones Collection

Clockwise from top left:

CRYSTAL BARBIE/WHITE   #4598

This doll is dressed in a white and iridescent fashion with a full ruffled boa. Her jewelry is a pair of crystal drop earrings and matching necklace. A child-size crystal style Barbie medallion is included in the package.

1984   NRFB $28.00

CRYSTAL KEN/BLACK   #9036

This Ken doll is dressed in an all-white outfit with a shiny vest and pink tie. His hair is molded and painted rather than rooted.

©1984   NRFB $25.00

CUTE 'N COOL BARBIE   #2954

This Target doll comes in a cotton knit outfit of a geometric pink, orange, purple, yellow, and white print. Five fashion separates plus extra accessories are included. There were a very limited number of boxes of this burgundy and gold style released. (The colors were chosen from the Dayton Hudson logo, the distributors for this doll.)

©1991   NRFB $43.00

DANCE CLUB BARBIE   #3509

This Children's Palace Special Edition doll includes a cassette with special Barbie dance music. Her outfit is a white vinyl jacket, black halter top, and hot pink skirt.

©1989   NRFB $48.00

Roxzella Jones Collection

Roxzella Jones Collection

Roxzella Jones Collection

Clockwise from top left:

DANCE MAGIC BARBIE   **#4836**
This doll is dressed in a gown that changes to a ballet outfit and to a disco outfit. Her lips will change colors from red to pink with hot or cold water from the enclosed applicator. Also included are a fan, ballet shoes, washcloth, cutouts, doll stand, poster, and more.
©1989   NRFB $24.00

DANCE MAGIC KEN   **#7081**
This doll is dressed in white, pink, and iridescent ballroom fashions. His hair changes color with the application of cold water.
©1990   NRFB $22.00

DAY-TO-NIGHT BARBIE/BLACK   **#7945**
This doll wears a double-breasted pink suit and has a pink party dress included.
©1984   NRFB $28.00

DAY-TO-NIGHT BARBIE/HISPANIC   **#7944**
This doll wears a double-breasted pink suit and has a pink party dress included. This Hispanic doll was produced in limited quantities.
©1984   NRFB $35.00

Roxzella Jones Collection

Clockwise from top left:

DAY-TO-NIGHT BARBIE/WHITE   **#7929**

This doll wears a double-breasted pink suit and has a pink party dress included. She came with an attaché case filled with magazines, calculator, business card, newspaper, and credit cards.

©1984   NRFB $30.00

DAY-TO-NIGHT KEN/BLACK   **#9018**

This doll wears a blue velvet suit, blue pin stripe slacks and vest and red and white tie. His hair is molded and painted and he also has a pink cummerbund and dark blue bow tie.

©1984   NRFB $22.00

DISNEY SPECIAL BARBIE   **#4385**

This is a Children's Palace Special Limited Edition. Barbie doll is dressed in a pink and white outfit with Minnie Mouse on her shirt. A blue denim jacket and black Mickey Mouse ears complete this outfit.

©1990   NRFB $50.00

DOCTOR BARBIE   **#3850**

Over 20 pieces are included with this Doctor Barbie doll. She wears a white lab coat over a pink dress and some of the items are stethoscope, doctor's bag, clipboard, and X-ray.

©1987   NRFB $50.00

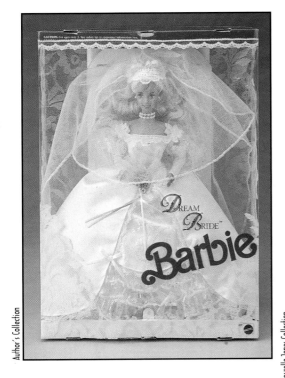

Author's Collection

Rozzella Jones Collection

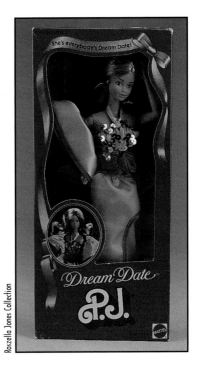

Rozella Jones Collection

Clockwise from top left:

DREAM BRIDE BARBIE   **#1623**
Barbie doll is dressed in a white satin floor-length gown with an iri-descent lace inset. This fashion has large, puffed sleeves. Her veil is attached to a lace headband.
©1991   NRFB $31.00

DREAM DATE BARBIE   **#5868**
This doll is wearing a unique fashion of magenta ruffles edged in pur-ple and a knit bodice accented with sequins.
©1983   NRFB $35.00

DREAM DATE P. J.   **#5869**
This doll is wearing a unique fashion of light blue ruffles edged in dark blue and a knit bodice accented with sequins.
©1983   NRFB $35.00

DREAM FANTASY BARBIE   **#7335**
A Wal-Mart Special Limited Edition doll, this Barbie doll is dressed in a pastel teal and silver bodysuit, tutu, and long skirt with extras.
1990   NRFB $45.00

Rozella Jones Collection

Roszella Jones Collection

Clockwise from top left:

DREAM GLOW BARBIE/BLACK   #2422
This pink gown is scattered with stars that will glow in the dark. This package includes a parasol, jewelry, glow-in-the-dark shoes and more.
©1985   NRFB $30.00

DREAM GLOW BARBIE/HISPANIC   #1647
This pink gown is scattered with stars that will glow in the dark. This package includes a parasol, jewelry, glow-in-the-dark shoes and more.
©1985   NRFB $43.00

DREAM GLOW BARBIE/WHITE   #2248
This pink gown is scattered with stars that will glow in the dark. This package includes a parasol, jewelry, glow-in-the-dark shoes and more.
©1985   NRFB $39.00

DREAM GLOW KEN/BLACK   #2421
Ken doll's hair is molded and painted and his vest and corsage will glow in the dark.
©1985   NRFB $23.00

Roszella Jones Collection

Clockwise from top right:

DREAMTIME BARBIE   #9180
This doll is wearing a pink, lavender, and silver nightgown and is holding a cuddly pink teddy bear with a silver bow. It is a Toys "Я" Us Special Edition.

1985	NRFB	$28.00

ENCHANTED EVENING BARBIE   #2702
This is a J. C. Penney Special Limited Edition doll dressed in a purple lame´ fashion with a metallic jeweltone and white fur-trimmed coat and fur hat.

©1991	NRFB	$112.00

EVENING FLAME BARBIE   #1865
This is a Special Limited Edition doll dressed in rich layers of red chiffon with a wrap of golden accented red tulle.

©1991	NRFB	$165.00

Roszella Jones Collection

Roszella Jones Collection

Clockwise from top left:

EVENING ENCHANTMENT BARBIE   #3596
This is a Sears Special Edition for 1989. She is dressed in a powder blue evening gown and her hair is pulled straight back with a blue ribbon that holds a mass of tightly curled light blonde hair.

©1989   NRFB $52.00

EVENING SPARKLE BARBIE   #3274
This doll is a Hills Special Limited Edition and is dressed in a white iridescent and blue evening fashion.

©1990   NRFB $40.00

FABULOUS FUR BARBIE   #7093
In this 4-piece set, Barbie doll wears a glittering blue jumpsuit with a white "fur" coat and white iridescent belt.

©1983   NRFB $50.00

FASHION JEANS BARBIE   #5315
This doll is dressed in jeans with pink trim and a fuzzy pink top.

1982   NRFB $38.00

Roszella Jones Collection

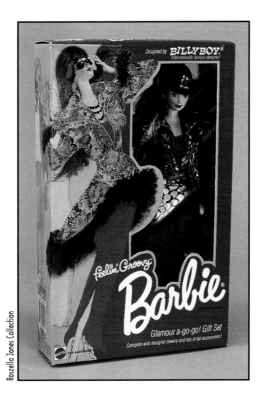

Roszella Jones Collection

Roszella Jones Collection

Clockwise from top left:

FASHION PLAY BARBIE #9629
This doll is dressed in a purple and white teddy and features the Barbie doll pink stamp advertising campaign. A membership form for the club is on the back of the box.
©1980   NRFB $18.00

FEELIN' GROOVY BARBIE #3421
This Limited Edition doll was designed by BillyBoy and includes sunglasses, shoulder bag, glittering shoes, travel case with luggage tag, hair dryer, camera, cutouts, and more. This is considered the first Barbie doll targeting the collectible market as compared to a doll for little girls' play. Le Nouveau Theatre Barbie doll was also designed by BillyBoy. Each is individually numbered and wears a black outfit with gold jewelry. The average value for Le Nouveau Theatre is $233.00
©1986   NRFB $189.00

FEELING FUN BARBIE #1189
This doll is dressed in a blue denim jacket and skirt trimmed in silver with a pink and silver lace overskirt. She is also wearing a white print blouse and pink boots.
©1988   NRFB $20.00

FOREIGN ANGEL FACE BARBIE #5640
This doll wears a very Victorian-looking outfit with a white lace top with high neck and puffed sleeves, black belt and satiny pink skirt. Her hair is unusual since it is sunstreaked. This doll was made for the foreign market and features foreign language on the box.
©1982   NRFB $40.00

Roszella Jones Collection

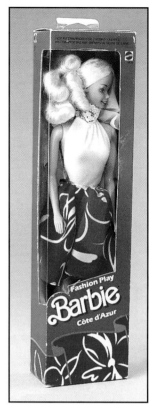

Rozella Jones Collection

Rozella Jones Collection

Rozella Jones Collection

Rozella Jones Collection

Clockwise from top left:

FOREIGN DIAMOND DREAM BARBIE
This foreign doll made for the Japanese market is dressed in a gown of pink ruffles with a white satin bodice. Her pink shoes have a special rhinestone accent as do her gown and earrings.

©1986   NRFB $86.00

FOREIGN DRESS ME BARBIE   #5696
This doll wears a white and purple teddy. This doll was made for the foreign market and features foreign language on the box.

©1990   NRFB $15.00

FOREIGN FASHION PLAY BARBIE   #4835
This foreign doll is dressed in a red and white satin skirt with a white halter top trimmed at the neck with lace. This doll was made for the foreign market and features foreign language on the box.

©1987   NRFB $22.00

FOREIGN FASHION PLAY BARBIE   #1380
This foreign doll is dressed in a teal dress with iridescent ruffle in the skirt. This doll was made for the foreign market and features foreign language on the box.

©1988   NRFB $20.00

Clockwise from top right:

FOREIGN FRUHLINGSZAUBER BARBIE   **#7546**

This is a German doll dressed in a long white, pink, purple, and silver gown with a ruffled wrap, straw hat, and basket. (A fashion package was also released of this gown. See Collector Series II in the fashion section.)

©1983   NRFB $78.00

FOREIGN HAPPY BIRTHDAY BARBIE   **#1922**

This doll's box features English, German, French, and Italian languages. She is wearing a dress with a bodice of layers of pastel blue, pink, and white lace. The skirt is white sheer material with a pastel underlayer.

©1980   NRFB $60.00

FOREIGN HAPPY BIRTHDAY BARBIE   **#9211**

This doll is dressed in a pink and iridescent gown with a full skirt that flows to the floor. The box is bilingual and includes a gift of a heart pendant.

©1989   NRFB $30.00

FOREIGN BARBIE IN INDIA   **#9910**

This special doll is dressed in a green satin outfit trimmed in red and gold brocade.

           NRFB $75.00

Roszella Jones Collection

Roszella Jones Collection

Roszella Jones Collection

Roszella Jones Collection

Clockwise from top left:

FOREIGN MY FIRST BARBIE #1875

This doll produced for the UK is dressed in a yellow dress with blue and red print. She comes with a free 40-page handbook.

©1980   NRFB $22.00

FOREIGN MY FIRST BARBIE #1875

This doll is dressed in a yellow bathing suit trimmed with blue. Also included are yellow pants, a blue skirt, a striped strapless top, and open-toe blue heels.

©1980   NRFB $22.00

FOREIGN MY FIRST BARBIE #1875

This foreign doll features a cotton gingham skirt and pink bodice edged with a white ruffle. The version for the American market is the same except for the box.

©1982   NRFB $20.00

FOREIGN MY FIRST BARBIE PRETTIEST PRINCESS EVER #9942

This foreign easy-to-dress doll wears a princess outfit with pink bodice complete with a white iridescent crown. The American market version has a purple bodice.

©1989   NRFB $16.00

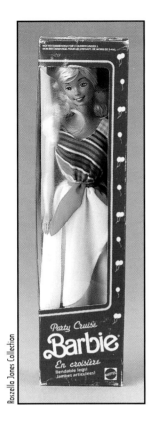

Clockwise from top left:

FOREIGN PARTY CRUISE BARBIE   #3075
A cruise outfit with white skirt and multi-colored top with turquoise
sash at the waist is featured on this doll.
©1986   NRFB $20.00

FOREIGN BARBIE PASSEIO   #10.5029
This foreign doll wears a beige jacket with wide red lapels and dark
leather belt. Also included are a red felt hat, brown leather purse, and
brown boots.
                               NRFB $88.00

FOREIGN RIVIERA BARBIE   #7344
This doll wears a pink, lime green, and black bathing suit.
©1989   NRFB $18.00

FOREIGN ST. TROPEZ BARBIE   #2096
Barbie doll is dressed in a magenta and black swim suit accented with
silver stars. This doll was released in Europe.
©1988   NRFB $28.00

Clockwise from top left:

FOREIGN SUNSATIONAL MALIBU BARBIE   #4970

This doll has that California look and is dressed in a blue bathing suit with purple trim. She has purple sunglasses. She was released in the Hispanic version for this foreign market production.

©1983   NRFB $28.00

FOREIGN FANTASY KEN

This Japanese Ken doll is dressed in a gray satin jacket, white satin shirt with pearl buttons and dark blue satin slacks. Included in the package are glasses, watch, shoes, and stand.

©1986   NRFB $86.00

FRIENDSHIP BARBIE   #5506

Issued in commemoration of the fall of the Berlin Wall, the package for this doll features the faces of children of numerous nationalities. She is wearing an off-the-shoulder pastel pink party dress trimmed with iridescent ruffles. She is packaged in a bilingual box and is referred to as the first Friendship Barbie doll.

©1990   NRFB $48.00

FRIENDSHIP BARBIE   #2080

Issued in commemoration of the fall of the Berlin Wall, the package for this doll features the faces of children of three nationalities. She is wearing a pink ruffled skirt, white shirt and black tank top with the Barbie "B" on the front. She is packaged in a bilingual box and is referred to as the second Friendship Barbie doll.

©1991   NRFB $48.00

Rozella Jones Collection

Rozella Jones Collection

Rozella Jones Collection

Clockwise from top right:

FRIENDSHIP BARBIE   #3677

Issued in commemoration of the fall of the Berlin Wall, the package for this doll features children of three nationalities holding hands. She is wearing a familiar red and white outfit trimmed in red hearts. She is packaged in a bilingual box and is referred to as the third Friendship Barbie doll.

©1991   NRFB $40.00

FRILLS & FANTASY BARBIE   #1374

The Wal-Mart 1989 Special Limited Edition doll wears a gown of layers of blue sheer material.

1989   NRFB $44.00

FUN TO DRESS BARBIE   #1372

This doll comes dressed in a light pink teddy trimmed with white and silver lace and offering a $2.00 rebate and a free poster.

©1988   NRFB $10.00

FUN TO DRESS BARBIE/BLACK   #4939

This doll comes dressed in a light pink top and panties trimmed in white lace.

©1989   NRFB $8.00

Roszella Jones Collection

Roszella Jones Collection

Roszella Jones Collection

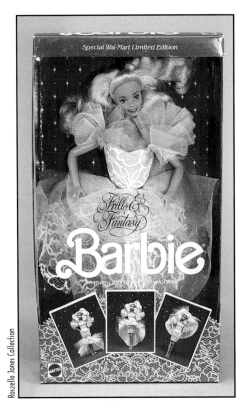

Roszella Jones Collection

Rozella Jones Collection

Rozella Jones Collection

Rozella Jones Collection

Rozella Jones Collection

Clockwise from top left:

FUNTIME BARBIE   **#1738**
This doll comes in a pink and silver crop top and shorts. This package includes a working watch and several small extras.
©1986   NRFB $20.00

GARDEN PARTY BARBIE   **#1953**
Barbie doll models a lacy dress with ruffled tulle sleeves. Included are a lacy overskirt, floral necklace, earrings, ring, shoes, hairbrush, and poster.
©1988   NRFB $20.00

GIFT GIVING BARBIE   **#1922**
Barbie doll comes with a surprise gift with lots of tiny extras. She is wearing a purple party dress and her hair is held back with a purple ribbon.
1986   NRFB $19.00

GIFT GIVING BARBIE   **#1205**
Barbie doll comes with a child's size gift necklace and lots of tiny extras. She is wearing a purple and white lace party dress with puffed purple iridescent sleeves.
©1988   NRFB $18.00

Roszella Jones Collection

Roszella Jones Collection

Roszella Jones Collection

Clockwise from top left:

GOLD & LACE BARBIE   #7476
A department store special for Target, this doll is dressed in a glitter-ing gold body suit, gold and white lace skirt and an overjacket sparkling with glitter.
1989   NRFB $40.00

GOLDEN DREAM BARBIE   #1874
This doll is dressed in a glittering gold jumpsuit with a sheer gold and white wraparound long skirt and cape.
©1980   NRFB $40.00

GOLDEN DREAM BARBIE   #1874
The first issue of this doll was shipped with an unusual hair style. The heads of some were exchanged at the retail store level and the replaced head was often discarded. Extra head $5.00.
©1980   NRFB $45.00

GOLDEN EVENING BARBIE   (on box) #2587
This doll is a Target exclusive and is dressed in black velvet straight skirt with gold and black bodice and a gold lamé jacket.
©1991   NRFB $38.00

Roszella Jones Collection

Rozella Jones Collection

GOLDEN GREETINGS BARBIE (F.A.O. SCHWARZ)   #7734
This doll is the first in a series of F.A.O. Schwarz Limited Edition dolls.
She is wearing a metallic gold evening fashion with a white lace and
gold-trimmed overskirt.
©1991   NRFB $235.00

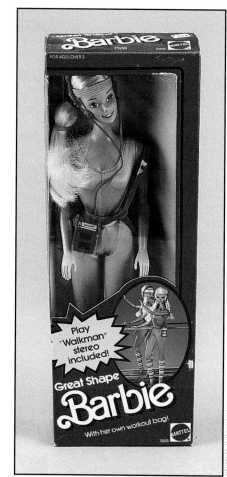

Rozella Jones Collection

GREAT SHAPE BARBIE/WHITE   #7025
Barbie doll wears a blue leotard with striped leg warmers. This package
is different because it also includes a play Walkman stereo.
©1983   NRFB $28.00

Roxzella Jones Collection

Roxzella Jones Collection

Clockwise from top left:

GREAT SHAPE BARBIE/BLACK   **#7834**
Barbie doll wears a blue leotard with striped leg warmers.
©1983   NRFB $17.00

GREAT SHAPE BARBIE/WHITE   **#7025**
Barbie doll wears a blue leotard with striped leg warmers. The Great
Shape Ken doll wears blue exercise pants, and a white T-shirt with a
yellow and blue horizontal stripe and tennis shoes. His average value
is $15.00.
©1983   NRFB $19.00

HAPPY BIRTHDAY BARBIE   **#1922**
This doll is dressed in a pink party dress with white flocked polka dots
and is holding a party gift. Her shoes are white open-toe heels.
1985   NRFB $16.00

Roxzella Jones Collection

HAPPY BIRTHDAY BARBIE #7913

This birthday doll is dressed in a pink gown with a full flowing skirt accented with hearts and confetti. She wears a Happy Birthday sash and her hair is accented with iridescent highlights. This birthday card was issued to coordinate with this doll.

©1990　　　　　NRFB　　　　$31.00

Rozella Jones Collection

Rozella Jones Collection

HAPPY BIRTHDAY BARBIE #3679

This birthday doll is dressed in a peach gown with a full flowing skirt accented with tiny pearls. She has a special birthday present included in the package.

©1991　　　　　NRFB　　　　$33.00

1988 HAPPY HOLIDAYS BARBIE   #1703
The first in the very successful Christmas Holiday series. This Barbie doll features a shimmering red gown.
1988   NRFB $625.00

1989 HAPPY HOLIDAYS BARBIE   #3523
The second in the Christmas Holiday series. This Barbie features a white gown trimmed in white "fur" with a glittering underskirt. A special silvery snowflake ornament is included in this Christmas package.
1989   NRFB $212.00

1990 HAPPY HOLIDAYS BARBIE/BLACK #4543
The third in the Christmas Holiday series. This Barbie doll features a
magenta gown with silver glittering starbursts and a starry ornament.
1990   NRFB $107.00

1990 HAPPY HOLIDAYS BARBIE/WHITE #4098
The third in the Christmas Holiday series. This Barbie doll features a
magenta gown with silver glittering starbursts and a starry ornament.
This doll is featured on the cover.
1990   NRFB $127.00

1991 HAPPY HOLIDAYS BARBIE/BLACK   #2696
The fourth in the Christmas Holiday series. This Barbie doll features
a rich green velvet gown trimmed in silvery sequins and red and
green beadwork.
1991   NRFB $100.00

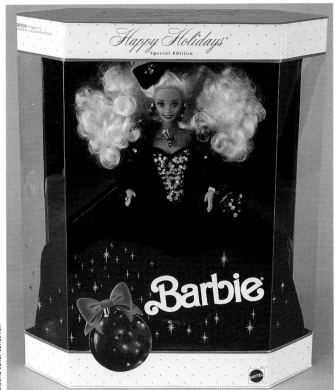

1991 HAPPY HOLIDAYS BARBIE/WHITE   #1871
The fourth in the Christmas Holiday series. This Barbie doll features
a rich green velvet gown trimmed in silvery sequins and red and
green beadwork.
1991   NRFB $138.00

Rozella Jones Collection

Rozella Jones Collection

Clockwise from top left:

HAWAIIAN BARBIE   **#7470**
This doll is dressed in a Hawaiian print two-piece bathing suit and wrap skirt. She has a band of flowers on her head.
1983   NRFB   **$38.00**

HAWAIIAN FUN BARBIE   **#5940**
This doll is dressed in a bright pink, lime, and yellow bikini. Included in the package are a hula skirt and sunglasses.
©1990   NRFB   **$19.00**

HAWAIIAN FUN BARBIE AND KEN   **#5940 & 5941**
These two Hawaiian Fun dolls were shrink wrapped together and sold as a Sam's Discount Club promotion.
©1990   NRFB   **$42.00**

HOLIDAY BARBIE   **#3406**
This was the second in the Applause series and is wearing a silver gown with a white and iridescent tulle overlay accented with pink.
1991   NRFB   **$46.00**

Clockwise from top right:

HOME PRETTY BARBIE   #2249

This doll wears an outfit that also features a lace-edged apron and other extras for many different fashion looks.

©1990   NRFB $20.00

HORSE LOVIN' BARBIE   #1757

The western doll comes dressed in red leather-look pants, red and white checkered shirt, tan leather-look vest trimmed in white "fur," cowboy hat and boots, and leather-look saddlebags.

©1982   NRFB $42.00

HORSE LOVIN' KEN   #3600

The western doll comes dressed in red leather-look pants, red and white checkered shirt, tan leather-look vest trimmed in white "fur," cowboy hat and boots, and leather-look saddlebags.

©1982   NRFB $32.00

HORSE LOVIN' SKIPPER   #5029

The western doll comes dressed in red leather-look pants, red and white checkered shirt, tan leather-look vest trimmed in white "fur," cowboy hat and boots, and lasso.

©1982   NRFB $32.00

Roszella Jones Collection

Roszella Jones Collection

Roszella Jones Collection

Roszella Jones Collection

Rozella Jones Collection

Rozella Jones Collection

Rozella Jones Collection

Rozella Jones Collection

Clockwise from top left:

HOT LOOKS BARBIE   **#5756**
This Special Edition for Ames is dressed in a pink print straight skirt outfit with dark blue gathered overskirt and blue leggings. She has matching fashions for 10 great looks!
©1991   NRFB $34.00

ICE CAPADES 50TH ANNIVERSARY BARBIE/BLACK   **#7348**
Made in the likeness of glamorous stars of the Ice Capades, this doll comes with a pink and purple practice suit and a dazzling show skirt. It is complete with white ice skates.
©1989   NRFB $25.00

ICE CAPADES 50TH ANNIVERSARY BARBIE/WHITE   **#7365**
Made in the likeness of glamorous stars of the Ice Capades, this doll comes with a pink and purple practice suit and a dazzling show skirt. It is complete with white ice skates.
©1989   NRFB $30.00

ICE CAPADES BARBIE/WHITE   **#9847**
This doll is wearing a purple, white, and iridescent skating outfit complete with white ice skates.
©1990   NRFB $27.00

Roszella Jones Collection

Roszella Jones Collection

Roszella Jones Collection

Clockwise from top left:

INTERNATIONAL BRAZILIAN BARBIE · #9094

Dolls of the World Collection. She is wearing a bare midriff outfit with pink and purple ruffled accents. The back of the box has important facts about the country and Portuguese words for you to learn.

©1989 · NRFB · $60.00

INTERNATIONAL CANADIAN BARBIE · #4928

Dolls of the World Collection. This special Canadian Barbie doll is dressed as a Royal Canadian Mounty in red velveteen jacket, black Mounty riding pants, round hat and boots.

©1987 · NRFB · $73.00

INTERNATIONAL CZECHOSLOVAKIAN BARBIE · #7330

Dolls of the World Collection. This doll has a very special white top with mini pleated sleeves, black bodice and white lace collar. Her skirt is bright yellow with a black and red print and white lace hem. Red tights and black lace-up boots complete this outfit. Made only one year, it is sometimes hard to find.

©1990 · NRFB · $105.00

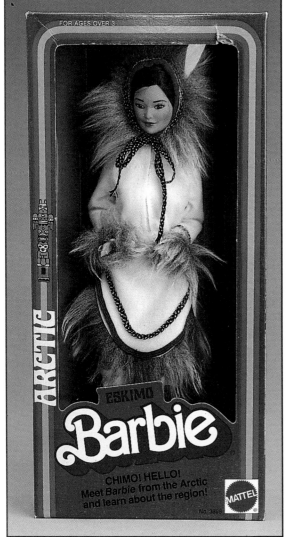

INTERNATIONAL (ARCTIC) ESKIMO BARBIE   #3898

This first International Arctic Eskimo doll is dressed in a white parka with fur, black, silver, and red trim. She wears white boots with red top stitching.

©1983   NRFB $107.00

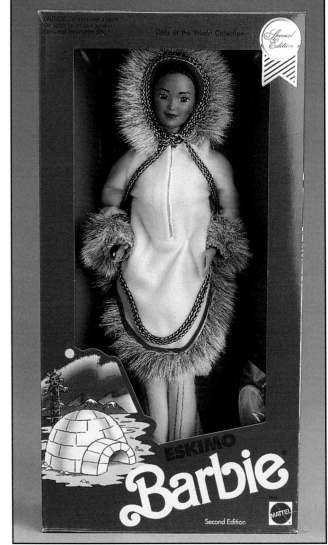

INTERNATIONAL (ARCTIC) ESKIMO BARBIE   #9844

Dolls of the World Collection. This Second Edition Eskimo doll is dressed in a white parka with fur, black, silver, and red trim. She wears white boots with red top stitching. Unless you have this and the first Eskimo doll side by side, it is sometimes hard to explain the difference between the two, especially out-of-box.

©1991   NRFB $50.00

INTERNATIONAL ENGLISH BARBIE #4973
Dolls of the World Collection. This doll wears a very stylish horseback riding outfit with a red velveteen jacket and dark blue skirt.
©1991 NRFB $59.00

INTERNATIONAL GERMAN BARBIE #3188
Dolls of the World Collection. This doll wears a beautiful red velvet waistcoat with dark printed skirt and white satin apron, topped with a wonderful blue and metallic gold bonnet.
©1986 NRFB $105.00

INTERNATIONAL GREEK BARBIE #2997
Dolls of the World Collection. This doll is dressed in bright red with black, gold, and white accents. She wears a red cap with gold tassel.
©1985   NRFB $85.00

INTERNATIONAL ICELANDIC BARBIE #3189
Dolls of the World Collection. This doll is dressed in a deep blue velvet jumper and white blouse. Her apron is light blue satin trimmed in metallic blue and gold.
©1986   NRFB $105.00

Roszella Jones Collection

INTERNATIONAL INDIA BARBIE   #3897
This International Barbie doll is dressed in a red satin outfit accented with brilliant golden trim and a golden underblouse.
©1983   NRFB $149.00

Roszella Jones Collection

INTERNATIONAL IRISH BARBIE   #7517
This International Barbie doll wears a satin Irish green outfit with a shawl trimmed in yellow. It has a white blouse trimmed with Irish lace and a white bonnet.
©1983   NRFB $130.00

INTERNATIONAL JAMAICAN BARBIE   #4647
Dolls of the World Collection. The back of the box has important facts about Jamaica and a map of the country. Some dolls have silver earrings and some blue. In some circles, the silver commands a higher price.
1991   NRFB $38.00

INTERNATIONAL JAPANESE BARBIE   #9481
Dolls of the World Collection. This doll comes with fan, stand, map, money, tour tickets, passport, and much more. She is dressed in a red satin Oriental print kimono with a golden obi.
©1984   NRFB $144.00

INTERNATIONAL KOREAN BARBIE   #4929

Dolls of the World Collection. This doll wears a bright magenta and lime green satin Korean outfit trimmed in gold.

©1987   NRFB $73.00

Rozella Jones Collection

INTERNATIONAL MALAYSIAN BARBIE   #7329

Dolls of the World Collection. This doll is dressed in a bright metallic magenta jacket with gold, black, and pink print skirt and trim.

©1990   NRFB $50.00

INTERNATIONAL MEXICAN BARBIE #1917
Dolls of the World Collection. This doll is wearing a flashy red satin skirt with a white blouse accented with brightly colored embroidery. The back of the box has important facts about the country and Spanish words for you to learn.
©1988 NRFB $50.00

Roszella Jones Collection

Roszella Jones Collection

INTERNATIONAL NIGERIAN BARBIE #7376
Dolls of the World Collection. This doll is wearing a unique fashion made from brown, black, and white animal-style print material. She has golden arm bands, neckpiece, and belt. The back of the box has important facts about Nigeria and a map of the country.
1989 NRFB $46.00

INTERNATIONAL ORIENTAL BARBIE   #3262

This International doll is dressed in a yellow dress and yellow and red jacket trimmed in metallic gold. A printed Oriental fan is included with this doll.

©1980   NRFB $150.00

INTERNATIONAL PARISIAN BARBIE   #9843

Dolls of the World Collection. This Second Edition Parisian doll wears a pink satin can-can outfit trimmed in black and silver. She has a pink cameo at her neck and wears black lace stockings. This is another doll that unless you have it and the first doll side by side, it is sometimes hard to explain the difference between the two, especially out-of-box.

©1990   NRFB $50.00

INTERNATIONAL PERUVIAN BARBIE   #2995
Dolls of the World Collection. This doll is dressed in a bright outfit with dark blue skirt accented with red, green, yellow, and white. Her blouse is dark pink and her cape is teal trimmed in dark pink.
©1985   NRFB $87.00

INTERNATIONAL RUSSIAN BARBIE   #1916
Dolls of the World Collection. This Russian doll wears a magenta velveteen dress accented by brown fur and gold trim. This doll is especially notable because it was released the year the Soviet government fell. This doll is also featured on the cover.
©1988   NRFB $78.00

INTERNATIONAL SCOTTISH BARBIE   #3263

This International doll wears a black blouse trimmed in white lace and a red plaid skirt and sash with a black Balmoral.
©1980   NRFB $138.00

Roszella Jones Collection

Roszella Jones Collection

INTERNATIONAL SCOTTISH BARBIE   #9845

Dolls of the World Collection. This Second Edition doll is dressed in a skirt and sash with a large plaid pattern. The material used in these Second Edition dolls has gold lamé accents in the plaid; the material used for the first did not.
©1990   NRFB $50.00

INTERNATIONAL SCOTTISH BARBIE #9845

Dolls of the World Collection. This Second Edition doll is dressed in a skirt and sash with a smaller plaid pattern than doll shown on previous page. The material used in these Second Edition dolls has gold lamé accents included in the plaid. The plaid material used for the first did not.

©1990 NRFB $50.00

Roszella Jones Collection

INTERNATIONAL SPANISH BARBIE #4031

This International doll is dressed in a red satin dress accented with black lace. She has a black lace fan and red flower for her hair.

©1982 NRFB $100.00

INTERNATIONAL SPANISH BARBIE #4963

Dolls of the World Collection. This ethnic doll is wearing a green satin skirt, black blouse and black lace head-scarf. She has a yellow shawl trimmed with black fringe and a red apron with yellow and green print. This doll is very easy to differentiate from the first Spanish issue because this outfit is completely new. ©1991 NRFB $45.00

INTERNATIONAL SWEDISH BARBIE #4032

This International doll is dressed in a dark blue velvet vest, lighter blue skirt with white apron and white lace-trimmed blouse. She wears a white bonnet and has a stand.
©1982 NRFB $109.00

Roszella Jones Collection

Roszella Jones Collection

Left to right:

INTERNATIONAL SWISS BARBIE   #7541
This International doll comes dressed in a white satin blouse and blue velvet vest top with a red velveteen printed skirt. Extras
include a straw hat, ski lift tickets, passport, money, and more.
©1983   NRFB $111.00

ISLAND FUN BARBIE   #4061
A tropical print wrap skirt, white bathing suit with a pink ruffle lei, and seahorse comb are included in this box.
©1987   NRFB $17.00

Roszella Jones Collection

SPECIAL EDITION

Jewel Jubilee
Barbie

Shine with beauty at
the season's most glamorous ball.

Roszella Jones Collection

Left to right:

ISLAND FUN CHRISTIE   **#4092**
A tropical print wrap skirt and yellow bathing suit with a pink ruffle lei, and seahorse comb are included in this box.
©1987   NRFB   **$15.00**

JEWEL JUBILEE BARBIE   **#2366**
The Second Limited Edition for wholesale club stores, this Barbie doll is dressed in white satin and tulle with glittering gold highlights and golden bows.
1991   NRFB   **$82.00**

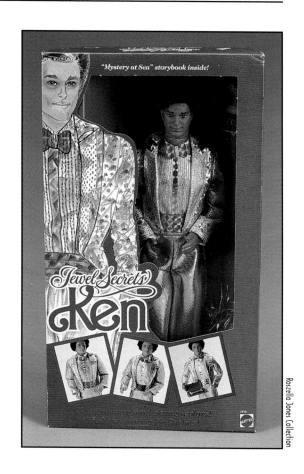

Roszella Jones Collection

Roszella Jones Collection

Clockwise from top left:

JEWEL SECRETS BARBIE #1737
A story book, "The Night of Jewel Secrets," is included with this Barbie doll glamorously dressed in sparkling silver and mauve satin gown and glittering shoes.
©1986 NRFB $27.00

JEWEL SECRETS KEN #1719
This doll used a face mold found also on the #3131-Hot Rockin' Ken doll. He is dressed in a silvery outfit with a blue cummerbund that changes colors. His hair is rooted.
©1986 NRFB $25.00

JEWEL SECRETS WHITNEY #3179
A story book, "The Night of Jewel Secrets," is included with this Whitney doll glamorously dressed in a sparkling silver and blue evening fashion. Styling extras are included to change her ensemble.
©1986 NRFB $29.00

Clockwise from top right:

LAVENDER LOOKS BARBIE   #3963
This is a Wal-Mart department store special. Included in this package are a minidress, skirt, gloves, shoes, brush, earrings, and ring.
1989   NRFB $44.00

LAVENDER SURPRISE BARBIE/BLACK   #5588
This is a Sears Special Edition doll. This doll wears a lavender ruffled minidress and includes an overskirt, flower on an elastic band, shoes, and brush.
©1989   NRFB $43.00

LAVENDER SURPRISE BARBIE/WHITE   #9049
This is a Sears Special Edition doll. This doll wears a lavender ruffled minidress and includes an overskirt, flower on an elastic band, shoes, and brush.
©1989   NRFB $47.00

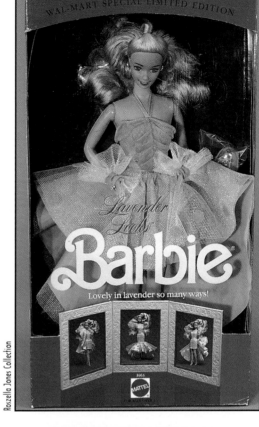

Roszella Jones Collection

Roszella Jones Collection

Roszella Jones Collection

Rozella Jones Collection

Rozella Jones Collection

Rozella Jones Collection

Clockwise from top left:

LIGHTS & LACE BARBIE   **#9725**
This Barbie doll is dressed in a hot pink lace outfit that features a light-up jewel and pink boots.
©1990   NRFP $21.00

LILAC & LOVELY BARBIE   **#7669**
This Barbie doll, as you might guess, is dressed entirely in lilac satin, lace, and tulle. She is a Sears Special Limited Edition.
©1987   NRFB $54.00

LOVING YOU BARBIE   **#7072**
This doll features the very recognizable red and white fashion with a red velvet bodice and white skirt with sheer white overskirt accented with flocked red hearts.
©1984   NRFB $25.00

MACKIE GOLDEN BARBIE #5405

This was the first in what has turned out to be one of the most popular Limited Edition series. Making his way into the doll fashions arena, Bob Mackie designed Golden Mackie Barbie doll even right down to the facial make up. This fashion has a bare midriff and utilizes an innumerable amount of gold sequins and beads. The downy feather boa is the perfect accent for this design. This doll is featured on the cover.

1990   Mint $822.00

Rozzella Jones Collection

MACKIE PLATINUM BARBIE   #2704

There seems to be some disagreement as to which of the next two dolls in Bob Mackie's series was second and which was third. Since they were released the same year and seem to be deliberately designed as opposites, it is safe to say these two were second and third in his series. This is a white doll with white hair. Her fashion is also all white. Her gown is entirely sequined with beads added at the neckline and bodice. Her overgarment has a high rounded collar and is accented with an abundance of sequins and beads.

1991   NRFB $675.00

Rozzella Jones Collection

MACKIE STARLIGHT SPLENDOR BARBIE   #2703

This design by Bob Mackie is a black doll with black hair and models a white and black sequined and beaded fashion with feather accents. This doll is reported to be hard to find.

1991   NRFB $700.00

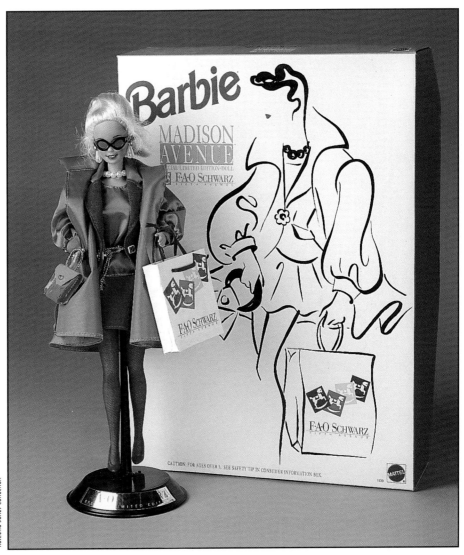

Roxella Jones Collection

Roxella Jones Collection

Top to bottom:

MADISON AVENUE BARBIE (F.A.O. SCHWARZ)   #1539
This stylish Barbie doll wears a lime green and magenta fashion with magenta stockings. She holds an F.A.O. shopping bag and comes with extras that make this doll a wonderful addition to your collection.
©1989   NRFB $188.00

MAGIC CURL BARBIE/BLACK   #3989
She is dressed in a bright yellow dress with fitted bodice and puffed sleeves. You could straighten or curl her hair with the use of the Magic Mist™ enclosed in the package.
©1981   NRFB $32.00

Rozella Jones Collection

Clockwise from top left:

MAGIC CURL BARBIE/WHITE   #3856
She is dressed in a bright yellow dress with fitted bodice and puffed sleeves. You could straighten or curl her hair with the use of the Magic Mist™ enclosed in the package.
©1981   NRFB $38.00

MAGIC MOVES BARBIE/BLACK   #2127
Barbie doll is dressed in a powder blue outfit trimmed in white "fur" tipped in light blue. Her arms will raise on their own when a switch on her back is pushed.
©1985   NRFB $28.00

MAGIC MOVES BARBIE/WHITE   #2126
Barbie doll is dressed in a powder blue outfit trimmed in white "fur" tipped in light blue. Her arms will raise on their own when a switch on her back is pushed.
©1985   NRFB $31.00

Rozella Jones Collection

Rozella Jones Collection

Roszella Jones Collection

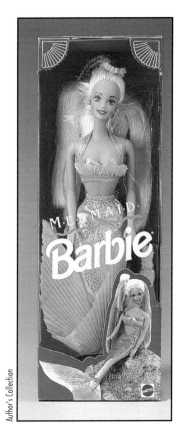

Author's Collection

Clockwise from top left:

MARDI GRAS BARBIE   **#4930**
The first of the American Beauties Collection, this special doll comes dressed in a gorgeous purple metallic gown complete with stockings. Included are a mask, purse, stand, and more.
©1987   NRFB $100.00

MERMAID BARBIE   **#1434**
This unusual doll dressed as a mermaid has hair that is iridescent and changes to a rainbow of colors when wet.
©1991   NRFB $25.00

MOONLIGHT ROSE BARBIE   **#3549**
This is a Hills Limited Edition doll. She is dressed in a form-fitting silver dress with a long overskirt of rose-colored tulle.
©1991   NRFB $40.00

Roszella Jones Collection

Rozella Jones Collection

Rozella Jones Collection

Rozella Jones Collection

Rozella Jones Collection

Clockwise from top left:

MY FIRST BARBIE   **#1875**
This second issue doll features a cotton gingham skirt and pink bodice edged with a white ruffle. The version for the foreign market is the same except for the box.
©1982   NRFB $18.00

MY FIRST BARBIE A GLITTERING BALLERINA   **#3839**
This doll is dressed in an iridescent blue tulle and satin ballerina costume and comes with blue toe shoes.
©1991   NRFB $10.00

MY FIRST BARBIE/BLACK   **#1801**
This doll has a twist 'n turn waist and is dressed in a pink ballet outfit with long tulle skirt and pink ballet slippers. Enclosed in the box is a yellow ballet ticket stub and instruction book.
©1986   NRFB $12.00

MY FIRST BARBIE/BLACK   **#1281**
This easy-to-dress Barbie doll comes in a white ballerina outfit complete with sparkling tutu and ballet shoes.
1988   NRFB $10.00

Clockwise from top left:

MY FIRST BARBIE PRETTIEST PRINCESS EVER/BLACK   #9944
This easy-to-dress doll wears a princess outfit with a purple bodice and iridescent skirt complete with a crown. This box features only Barbie doll printed on the front.
©1989   NRFB $10.00

MY FIRST BARBIE PRETTIEST PRINCESS EVER/WHITE   #9942
This easy-to-dress doll wears a princess outfit with a purple bodice and iridescent skirt complete with a crown. This box features Barbie and Ken dolls printed on the front. The foreign market version has a pink bodice.
©1989   NRFB $12.00

MY FIRST BARBIE/WHITE   #1280
This easy-to-dress Barbie doll comes in a white ballerina outfit complete with sparkling tutu and ballet shoes.
1988   NRFB $12.00

MY FIRST KEN   #9940
This easy-to-dress doll wears a purple satin jacket and comes complete with a crown. He was made to escort the My First Barbie Prettiest Princess Ever doll.
©1989   NRFB $12.00

Roszella Jones Collection

Roszella Jones Collection

Top to bottom:

NIGHT SENSATION BARBIE (F.A.O. SCHWARZ) #2921
This doll is the third in a series of F.A.O. Schwarz Limited Edition dolls. She is wearing an elegant evening gown of black satin with a fuchsia exaggerated ruffle on the bodice. This doll is featured on the cover.
©1990   NRFB $200.00

NURSE WHITNEY #4405
Over 20 pieces are included with this Nurse Whitney doll. She wears a white nurse's uniform and cap and some of the items are stethoscope, doctor's bag, clipboard, X-ray, blood pressure gauge and cord, bottles, tray, and more.
©1987   NRFB $79.00

Rozella Jones Collection

Rozella Jones Collection

Rozella Jones Collection

Clockwise from top left:

PARTY IN PINK BARBIE #2909
This is a Special Edition for Ames and she is dressed in a hot pink minidress with black collar and black lace sleeves. Extras include a silver and black overskirt and black leggings and crop top.
©1991   NRFB $32.00

PARTY LACE BARBIE #4843
This doll is a Special Limited Edition for Hills Department Stores. She is wearing a lavender off-one-shoulder party dress with lace at the top and at the hem.
©1989   NRFB $35.00

PARTY PINK BARBIE #7637
This Special Edition doll for Winn-Dixie comes in a pink and silver party dress.
©1989   NRFB $20.00

PARTY PRETTY BARBIE #5955
A Target Exclusive, this doll is dressed in a black party dress trimmed in white and iridescent lace and a jacket made from this lace material.
©1990   NRFB $34.00

Rozella Jones Collection

Clockwise from top left:

PARTY SENSATION BARBIE   **#9025**
This wholesale club doll is dressed in a party gown of layers of hot pink and red with detached puffed sleeves. Also included are hair ornament, pendant necklace, and more.
©1990   **NRFB**   **$54.00**

PARTY TREATS BARBIE   **#4885**
A pink party dress with iridescent overlay gives this Toys "Я" Us Barbie doll a special look.
©1989   **NRFB**   **$27.00**

PEACH PRETTY BARBIE   **#4870**
This is a K-Mart department store special and is dressed in a silver and peach gown.
©1989   **NRFB**   **$46.00**

PEACHES 'N CREAM BARBIE/BLACK   **#9516**
A sheer peach gathered skirt with ruffle at the bottom and white iridescent bodice are the fashion for this doll.
©1984   **NRFB**   **$22.00**

Clockwise from top right:

PEACHES 'N CREAM BARBIE/WHITE   #7926
A sheer peach gathered skirt with ruffle at the bottom and white sparkling bodice and matching stole are the fashion for this doll.
©1984   NRFB $30.00

PEPSI SPIRIT BARBIE   #4869
Toys "Я" Us, Pepsi and Mattel released this special Barbie doll complete with several outfits, beach bag, beach towel, headband, and more.
©1989   NRFB $80.00

PEPSI SPIRIT SKIPPER   #4867
Toys "Я" Us, Pepsi and Mattel released this special Skipper doll complete with several outfits, beach bag, beach towel, bathing suit, and more.
©1989   NRFB $80.00

Roszella Jones Collection

Roszella Jones Collection

Roszella Jones Collection

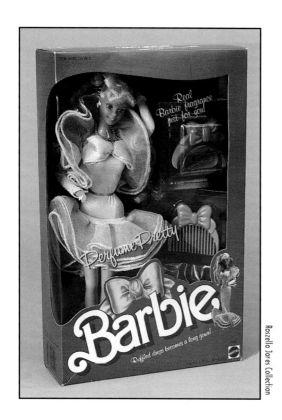

Roszella Jones Collection

Clockwise from top left:

PERFUME PRETTY BARBIE #4551
Barbie doll is dressed in pink tulle ruffles and the extra in this package is a half-ounce bottle of perfume.
©1987 NRFB $26.00

PINK JUBILEE BARBIE #4589
This is the Wal-Mart 25th anniversary doll wearing a pink minidress with a tulip skirt and topped with a pink cape. (Not to be confused with the scarce Pink Jubilee doll given at the 30th Anniversary for Barbie at the Lincoln Center in New York.)
1988 NRFB $73.00

PINK & PRETTY BARBIE #3554
This Barbie doll comes with a silky pink and glittery silver fashion with a pink "fur" wrap, and extra pieces for many dreamy looks.
©1982 NRFB $36.00

Roszella Jones Collection

Roszella Jones Collection

Roszella Jones Collection

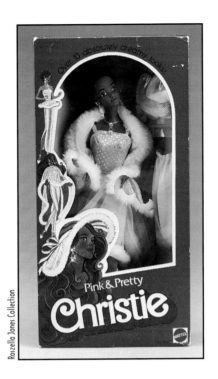

Roszella Jones Collection

Roszella Jones Collection

Clockwise from top left:

PINK & PRETTY BARBIE–EXTRA SPECIAL   #5239
This Barbie doll comes with a silky pink and glittery silver fashion with a pink "fur" wrap, and an extra outfit. Also included in this special package is a modeling set with TV camera, stage light, pink sunglasses, autograph book, and more.
©1981   NRFB $78.00

PINK & PRETTY CHRISTIE   #3555
This Christie doll comes with a silky pink and glittery silver fashion with a pink "fur" wrap, and extra pieces for many dreamy looks.
©1982   NRFB $30.00

PINK SENSATION BARBIE   #5410
This special Winn-Dixie Limited Edition of Barbie doll comes in a pink and white party dress.
1990   NRFB $20.00

PORCELAIN BENEFIT PERFORMANCE   #5475
This porcelain doll is wearing a red and white evening fashion. Her tiny earrings and necklace add a special touch to this doll.
1988   NRFB $433.00

Roszella Jones Collection

Clockwise from top left:

PORCELAIN BLUE RHAPSODY   #1708

This porcelain doll is wearing an elegant evening gown featuring a black gathered floor length skirt with an overlay skirt of sheer dark blue accented with gold and blue glitter. It has a blue satin bodice and sleeves. Her finely crafted blue jewel necklace and earrings complete this special ensemble. This doll is featured on the cover.

1986	NRFB	$740.00

PORCELAIN ENCHANTED EVENING   #3415

This blonde porcelain doll is wearing a replica of the 900 Series fashion by the same name.

1987	NRFB	$458.00

PORCELAIN GAY PARISIENNE   #9973

This brunette porcelain doll is wearing a replica of the 900 Series fashion by the same name.

1991	NRFB	$276.00

Roszella Jones Collection

Roszella Jones Collection

Clockwise from top left:

PORCELAIN KEN (GROOM)   **#1110**
This porcelain Ken doll is stylishly dressed as the groom in black and white.
1991   NRFB $208.00

PORCELAIN WEDDING DAY   **#2621**
This porcelain Barbie doll is wearing a beautiful white satin and lace full-length wedding gown and white tulle veil. The veil is held to her head with a halo of pearls and she is wearing a pearl necklace and coordinating earrings. It is a replica of the 900 Series fashion by the same name.
1989   Mint $590.00

Roszella Jones Collection

PORCELAIN SOLO IN THE SPOTLIGHT   #7613
This blonde porcelain doll is wearing a replica of the 900 Series fashion by the same name.
1990   NRFB $285.00

PORCELAIN SOPHISTICATED LADY   #5313
This brunette porcelain doll is wearing a replica of the 900 Series fashion by the same name.
1990   NRFB $270.00

Roszella Jones Collection

Roszella Jones Collection

Roszella Jones Collection

Clockwise from top left:

PRETTY CHANGES BARBIE   #2598

This doll is dressed in a bright yellow satin jumpsuit with a wraparound sheer yellow and white long skirt. You could change her looks with the enclosed picture hat, and blonde and brunette falls in the package. A star-shaped stand is included.

©1978   NRFB $35.00

PRETTY HEARTS BARBIE   #2901

This doll was a special edition for supermarkets and features the very recognizable party dress of red velvet bodice, white skirt with sheer white overskirt accented with flocked red hearts. The top has puffed sleeves of that same material.

©1991   NRFB $26.00

RAPPIN' ROCKIN' BARBIE   #3248

This doll is in an outfit of a black leather miniskirt and jacket, pink shirt, and comes with a real rap beat boom box.

©1991   NRFB $36.00

ROCKER BARBIE   #1140

This Rocker Barbie variation is dressed in a silver bodice with purple belt, white miniskirt and pink hair bow. Her tights are pink with horizontal silver stripes. The package includes iron-on decals and cassette with 4 songs. The box background on this particular doll is lime green and black.

©1985   NRFB $45.00

Roszella Jones Collection

Rozella Jones Collection

Rozella Jones Collection

Rozella Jones Collection

Rozella Jones Collection

Clockwise from top left:

ROCKER BARBIE #1140
This Rocker Barbie variation is dressed in a silver bodice with purple belt, white miniskirt and pink hair bow. Her tights are pink and silver overall. The package includes iron-on decals and cassette with 4 songs. The box background on this particular doll is bright yellow and the front of the box is slightly different from Rocker Barbie doll on previous page.

©1985   NRFB $45.00

ROCKER BARBIE #3055
Second Edition Barbie doll wears silver pants, pink miniskirt and top accented with silver stars and the Rocker logo. This doll comes with dancing action and a pink background in the box.

©1986   NRFB $38.00

ROCKER BARBIE #3055
Second Edition variation (possibly Canadian) Barbie doll wears silver pants, pink miniskirt and top accented with silver stars and the Rocker logo. This doll comes with dancing action and a blue background in the box.

©1986   NRFB $38.00

ROCKER DANA #1196
Dana doll wears blue leggings, red tank top and white jacket. The box background is pink.

©1985   NRFB $42.00

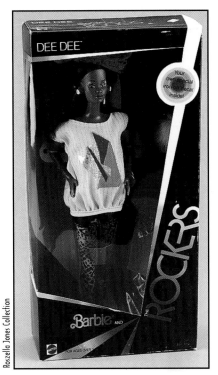

Clockwise from top left:

ROCKER DANA   #1196
First edition Rocker Dana variation wears blue leggings, red tank top
and white jacket. The box is slightly different from the previous page
Dana.
©1985   NRFB $42.00

ROCKER DEE DEE   #1141
Dee Dee doll wears multicolored tights and a yellow pullover top with
geometric art on the front and a leather miniskirt. The box background is bright orange.
©1985   NRFB $42.00

ROCKER DEE DEE   #1141
This Dee Dee doll variation wears multicolored tights and a lime
green and metallic gold striped pullover top with geometric art on
the front and a black leather miniskirt. The box background is lime
green and black and the front of the box is slightly different.
©1985   NRFB $42.00

ROCKER DEREK   #2428
Derek doll wears a one piece black slacks and yellow shirt outfit. A
black tie and a pink, blue, and silver jacket complete his outfit. This
unique face mold for Derek doll was used only in 1986.
1986   NRFB $60.00

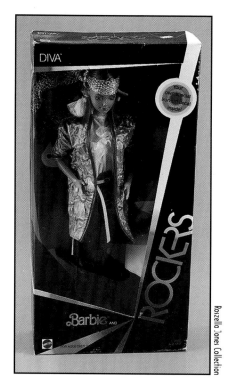

Roszella Jones Collection

Roszella Jones Collection

Roszella Jones Collection

Roszella Jones Collection

Clockwise from top left:

ROCKER DIVA   **#2427**
First Edition Rocker Diva variation wears fuchsia metallic pants and yellow shirt. A blue metallic jacket and silver headband complete this outfit. This box is slightly different with a lime green and black background.
©1985   NRFB $42.00

ROCKER DIVA   **#2427**
This Rocker Diva doll wears fuchsia metallic pants and yellow shirt. A blue metallic jacket and silver headband complete this outfit. The box has a bright yellow background and the front of the box is slightly different.
1986   NRFB $42.00

ROCKER KEN   **#3131**
Second Edition Rocker Ken doll is dressed in an outfit with a long silver overcoat. This doll's face mold was also used for #1719 Jewel Secrets Ken doll.
©1986   NRFB $60.00

ROLLER BLADE BARBIE   **#2214**
Dressed in white vinyl shorts and crop top, this doll wears special pink roller blades. This doll was temporarily recalled because of a fire danger resulting from the spark the rollers would emit when used.
©1991   NRFB $30.00

Roxzella Jones Collection

Roxzella Jones Collection

Roxzella Jones Collection

Clockwise from top left:

ROLLER SKATING BARBIE   #1880
This doll is dressed in purple shirt and black shorts trimmed in metallic red and purple. Included are white roller skates with red rollers.
©1980   NRFB $48.00

ROLLER SKATING KEN   #1881
This doll is dressed in purple shirt and black shorts trimmed in metallic red and purple. Included are white roller skates with red rollers.
©1980   NRFB $40.00

SCHOOL FUN BARBIE   #2721
This Toys "Я" Us doll wears a hot pink letter jacket and white print skirt. The package contains 2 pencils, a backpack, and more.
©1991   NRFB $26.00

BARBIE SENSATIONS BARBIE   #4931
A fifties theme was featured in this group of dolls. Barbie doll wears a pink and silver dropwaist dress with a full skirt and white and pink jacket both accented with musical notes. Belinda ($40.00), Bobby ($56.00), and Bopsy ($40.00) were dolls also produced for the Sensations theme.
©1987   NRFB $38.00

Roxzella Jones Collection

Roszella Jones Collection

Roszella Jones Collection

Roszella Jones Collection

Roszella Jones Collection

Clockwise from top left:

SHOW 'N RIDE BARBIE   **#7799**
This Toys "Я" Us package contains Barbie doll in an equestrian outfit plus a long skirt, riding crop, riding blanket, first prize ribbon, 4 horse leg warmers, 4 horseshoes, and more!
©1988   NRFB $40.00

SINGAPORE BARBIE
This Barbie doll was produced for the Singapore Airlines and has a wonderful bubble cut hairstyle. The package contains the doll, blouse, sarong, sandals, panties, hairbrush, and doll stand.
©1991   NRFB $120.00

SKATING STAR BARBIE   **#4547**
This doll is dressed in an all white skating outfit and holds a plastic bouquet of flowers. White ice skates are included. This was an official licensed product for the 1988 Calgary Winter Olympic Games.
©1987   NRFB $48.00

SKI FUN BARBIE   **#7511**
Barbie doll is dressed in a ski outfit of gold striped hot pink pants and bright pink, orange, blue, yellow, and green, gold-glittered jacket. Skis, ski poles, and more are included with this doll. This box is bilingual.
©1990   NRFB $18.00

Clockwise from top right:

SKI FUN KEN   #7512

Ken doll is dressed in a ski outfit of gold striped black, white, and lime green pants and matching jacket. Skis, ski poles, and more are included with this doll. This box is bilingual.

©1990   NRFB $20.00

SKI FUN MIDGE   #7513

Midge doll is dressed in a ski outfit of lime striped blue pants and blue fur trimmed jacket. Skis, ski poles and more are included with this doll. This box is bilingual.

©1990   NRFB $28.00

SOUTHERN BEAUTY BARBIE   #3284

This Winn-Dixie Special Edition doll is dressed in a peach satin party dress trimmed in white iridescent lace.

©1991   NRFB $29.00

SOUTHERN BELLE BARBIE   #2586

A Sears Special Edition, this Southern Belle doll is dressed in layers of peach ruffles and carries a purple umbrella.

©1991   NRFB $50.00

Roszella Jones Collection

Roszella Jones Collection

Roszella Jones Collection

Roszella Jones Collection

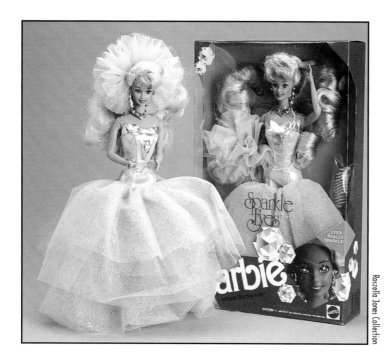

Rozella Jones Collection

Rozella Jones Collection

Rozella Jones Collection

Clockwise from top left:

SPARKLE EYES BARBIE #2482

This doll has special blue eyes that really sparkle. She is dressed in a pink, silver, and iridescent gown that will adapt to two other styles.

©1991 NRFB $23.00

SPECIAL EXPRESSIONS BARBIE #4842

This special Limited Edition doll for Woolworth is wearing a white satin sheath with a short silver and white overskirt.

©1989 NRFB $23.00

SPECIAL EXPRESSIONS BARBIE #5504

This special Limited Edition doll for Woolworth is wearing a pink sheath with tulle overskirt and trim.

©1990 NRFB $21.00

SPECIAL EXPRESSIONS BARBIE #2582

This special Limited Edition doll for Woolworth is wearing a teal party dress with shimmery lace accents.

©1991 NRFB $20.00

Rozella Jones Collection

Clockwise from top left:

SPRING PARADE BARBIE   **#7008**

This is a Toys "Я" Us Limited Edition doll perfect for Easter. She is dressed in a gown of lavender tulle with an iridescent fitted bodice, and a white wide brimmed hat. She is carrying a basket of flowers on her arm.

©1991   NRFB $39.00

STAR DREAM BARBIE   **#4550**

A Sears Special Limited Edition, this doll is dressed in a white tulle gown, glittery belt, and sparkly stockings with a bouquet of roses.

©1987   NRFB $72.00

STARS 'N STRIPES AIR FORCE BARBIE   **#3360**

This Stars 'n Stripes Special Edition Air Force Barbie doll is dressed in a flight suit and jacket, boots, cap, and scarf. The jacket is a reproduction of the official A-2 leather flight jacket issued to mission qualified aircrew members.

1991   NRFB $53.00

STARS 'N STRIPES MARINE CORPS BARBIE   **#7549**

In this Stars 'n Stripes Special Edition, Barbie doll is dressed in authentic Marine Corps "dress blues" and features a multi-stripe Desert Storm medal.

1991   NRFB $26.00

Roszella Jones Collection

Roszella Jones Collection

Roszella Jones Collection

Roszella Jones Collection

Roszella Jones Collection

Roszella Jones Collection

Roszella Jones Collection

Clockwise from top left:

STARS 'N STRIPES NAVY BARBIE   #9693
This Stars 'n Stripes Special Edition Navy Barbie doll is dressed in a white Navy uniform and comes with extras including a nautical map, cutouts, trousers, and extra shoes.
©1990   NRFB $27.00

STERLING WISHES BARBIE   #3347
This Spiegel Limited Edition doll is wearing an exquisite black and silver off-the-shoulder ballgown.
©1991   NRFB $170.00

STYLE MAGIC BARBIE   #1283
This doll introduced WondraCurl hair that would style easily and hold a curl. She is dressed in a pink party dress and comes with curling wand, haircomb, cross-sell poster, and instructions.
©1988   NRFB $25.00

STYLE MAGIC CHRISTIE   #1288
This doll introduced WondraCurl hair that would style easily and hold a curl. She is dressed in a peach party dress and comes with curling wand, haircomb, cross-sell poster, and instructions.
©1988   NRFB $22.00

Roxella Jones Collection

Roxella Jones Collection

Clockwise from top left:

SUMMIT BARBIE/ASIAN	#7029	©1990	NRFB	$32.00
SUMMIT BARBIE/BLACK	#7028	©1990	NRFB	$26.00

These Summit dolls were to commemorate the first annual Barbie Summit in 1990 with children from 30 countries participating in this unique cultural exchange. These dolls were intended as a symbol to remind us all that the children of the world can make a world of difference.

SUMMIT BARBIE/HISPANIC	#7030	©1990	NRFB	$32.00
SUMMIT BARBIE/WHITE	#7027	©1990	NRFB	$26.00

Roxella Jones Collection

Roxella Jones Collection

Rozella Jones Collection

Rozella Jones Collection

Rozella Jones Collection

Left to right:

SUN GOLD MALIBU BARBIE/BLACK   #7745
This summer Barbie doll comes in a gold and metallic checked one-piece swimsuit.
©1983   NRFB $15.00

SUN GOLD MALIBU BARBIE/WHITE   #1067
This summer Barbie doll comes in a gold and metallic checked one-piece swimsuit.
©1983   NRFB $18.00

SUN GOLD MALIBU KEN/BLACK   #3849
This Ken doll has molded painted hair and is dressed in teal trunks with gold accents.
©1983   NRFB $15.00

Clockwise from top left:

SUNSATIONAL MALIBU BARBIE/WHITE   #1067
This doll has that California look and is dressed in a purple bathing suit and has purple sunglasses.

©1981   NRFB $27.00

SUNSATIONAL MALIBU P. J.   #1187
This doll has that California look and is dressed in a teal bathing suit and has purple sunglasses. Her pigtails are bound with two strands of multicolored beads.

©1981   NRFB $26.00

SUNSATIONAL MALIBU KEN/BLACK   #3849
This doll has rooted hair and is dressed in yellow trunks.

©1981   NRFB $16.00

SUNSATIONAL MALIBU CHRISTIE   #1187
This doll has that California look and is dressed in a yellow bathing suit with red trim and has purple sunglasses. Her pigtails are bound with two rubber bands.

©1981   NRFB $23.00

Clockwise from top left:

SUPER HAIR BARBIE/BLACK   #3296

This doll comes with a unique hairstyling barrette and is dressed in a white and silver jumpsuit.

©1986   NRFB $20.00

SUPER HAIR BARBIE/WHITE   #3101

This doll comes with a unique hairstyling barrette and is dressed in a white and silver jumpsuit.

©1986   NRFB $23.00

SUPERSTAR BARBIE   #1604

She comes dressed in a pink gown sprinkled with silver stars. The package includes child-size charm bracelet with a star charm.

1988   NRFB $28.00

SUPERSTAR KEN/BLACK   #1550

Ken doll is brilliantly dressed in a silver tux with white satin slacks and silver starred white shirt. A silvery award for Barbie doll is included in this package. His hair is molded and painted and his face has the added touch of a mustache.

©1988   NRFB $22.00

Roszella Jones Collection

Roszella Jones Collection

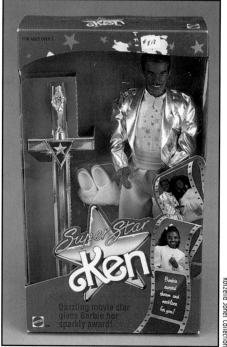

Roszella Jones Collection

Roszella Jones Collection

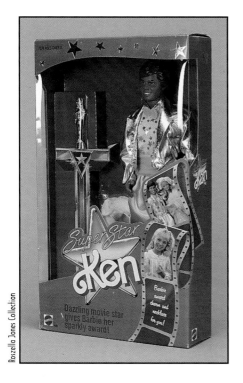

Roxzella Jones Collection

Roxzella Jones Collection

Roxzella Jones Collection

Clockwise from top left:

SUPERSTAR KEN/WHITE #1535
Ken doll is brilliantly dressed in a silver tux with white satin slacks and silver starred white shirt. A silvery award for Barbie is included in this package. A new Ken face mold was introduced on this doll.
©1988   NRFB $25.00

SWAN LAKE BARBIE #1648
This elegant ballerina is the first in the Prima Ballerina Series and comes in a circular display case and is standing on a rotating musical base. The Swan Queen from Tchaikovsky's ballet is the focus for this doll.
1991   NRFB $184.00

SWEET ROMANCE BARBIE #2917
This is a Limited Edition doll from Toys "Я" Us and Barbie is dressed in a blue tulle gown with blue metallic bodice and sleeves. A fragrance locket is included in this package.
©1991   NRFB $18.00

Clockwise from top left:

SWEET ROSES BARBIE #7635
With the increasing popularity of the Sweet Roses line of Barbie furniture, this Toys "Я" Us doll was renamed from the previous year's Home Pretty Barbie doll.
©1991 NRFB $39.00

SWEET ROSES P. J. #7455
This P. J. doll is dressed in a strapless gown with a layered skirt of pink satin. She is rose-scented and trimmed with ribbons and fabric roses.
©1983 NRFB $44.00

SWEET SPRING BARBIE #3208
Dressed in a bright dress of yellow, blue, white, and magenta with sheer magenta ruffles, this is a Special Edition Doll.
1991 NRFB $26.00

TEEN TALK BARBIE #5745
Barbie doll is dressed in a blue and white polka dot jacket and multicolored skirt. When a button on her back is pressed, she will say an array of phrases. A mistaken switch in recordings in the first issue of this doll makes her more valuable (up to $150.00), but the box must be opened to find out which you have. To some, opening the box devalues the item. So you must decide what to do!
1991 NRFB $45.00

Rozella Jones Collection

Rozella Jones Collection

Rozella Jones Collection

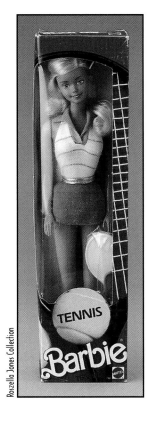

Roszella Jones Collection

Clockwise from top left:

TENNIS BARBIE   **#1760**
Pink tennis shoes with white socks trimmed in pink finish this halter-top pink, white, and gold tennis outfit. Racket and brush complete the extras with this doll.
©1986   NRFB $22.00

TOTALLY HAIR BARBIE/BLONDE   **#1112**
This doll is advertised to have the longest hair ever and comes dressed in a print minidress of pink, fuchsia, teal, purple, and white. A bottle of Dep styling gel is included in this package along with extra hair accessories.
©1991   NRFB $29.00

TOTALLY HAIR BARBIE/BRUNETTE   **#1117**
This doll is advertised to have the longest hair ever and comes dressed in a print minidress of hot pink, 2 shades of blue, purple, lime green, and white. A bottle of Dep styling gel is included in this package along with extra hair accessories.
©1991   NRFB $35.00

BRIDAL TRACY   **#4103**
This Tracy doll is dressed in a white satin wedding dress with lace sleeves and bodice with a high-neck collar and white tulle veil. She is holding a bouquet of plastic flowers.
1983   NRFB $50.00

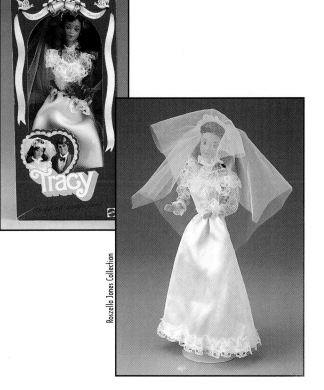

Roszella Jones Collection

Rozella Jones Collection

Rozella Jones Collection

Rozella Jones Collection

Clockwise from top left:

TODD #4253

Todd doll is the groom doll for Bridal Tracy doll and is dressed in a purple velvet and satin tuxedo with gray slacks and white ruffled front shirt.
©1982　　　　　　NRFB　　　　　　$50.00

TRAILBLAZIN' BARBIE #2783

Advertised as "steppin' lively" in cowgirl-cute Western wear, this is a Special Edition Doll complete with red Western boots.
1991　　　　　　NRFB　　　　　　$30.00

DELUXE TROPICAL BARBIE #2996

A swimsuit for each doll in this line is made from material with a wild tropical print on a black background. Included in this deluxe package are a wrap skirt, beach bag and straw hat, towel, sunglasses, swimfins, mask, snorkel, magazines, camera, and surfboard.
©1985　　　　　　NRFB　　　　　　$38.00

TROPICAL BARBIE/BLACK #1022

A swimsuit for each doll in this line is made from material with a wild tropical print on a black background.
©1985　　　　　　NRFB　　　　　　$14.00

Rozella Jones Collection

Clockwise from top right:

TROPICAL KEN/BLACK   #1023
A swimsuit for each doll in this line is made from material with a
wild tropical print on a black background.
©1985   NRFB $16.00

TWIRLY CURLS BARBIE/HISPANIC   #5724
This Department Store Special doll was sold one year only. She is dressed
in a pink outfit with a silver belt; this package includes the Twirly Curler.
©1982   NRFB $36.00

TWIRLY CURLS BARBIE/BLACK   #5723
This Department Store Special doll was sold one year only. She is dressed
in a pink outfit with a silver belt; this package includes the Twirly Curler.
©1982   NRFB $26.00

TWIRLY CURLS BARBIE/WHITE   #5579
This Department Store Special doll was sold one year only. She is dressed
in a pink outfit with a silver belt; this package includes the Twirly Curler.
©1982   NRFB $30.00

Rozella Jones Collection

Rozella Jones Collection

Rozella Jones Collection

Rozella Jones Collection

Clockwise from top left:

UNICEF BARBIE/ASIAN   #4774   1989   NRFB $35.00   UNICEF BARBIE/BLACK   #4770   1989   NRFB $30.00

A portion (37¢) of the sale price for each of these Special Edition dolls went to the U.S. Committee for UNICEF. The gown for these dolls has a gathered glittery dark blue floor length skirt with a sparkling white strapless fitted bodice and fitted sleeves. A red sash and special necklace complete the look. Since this was a promotion for UNICEF, the dolls come in Asian, Black, Hispanic, and White.

UNICEF BARBIE/WHITE   #1920   1989   NRFB $30.00   UNICEF BARBIE/HISPANIC   #4782   1989   NRFB $35.00

Clockwise from top right:

UNITED COLORS OF BENETTON BARBIE #9404

This doll is dressed in a sensational Benetton outfit with a red jacket trimmed in blue, yellow blouse, red miniskirt, blue print leggings, pink leg warmers, red shoes, and red felt hat. There are several dolls in this series.

©1990 NRFB $33.00

VACATION SENSATION BARBIE #1675

Barbie doll travels with sportswear and accessories including a pink and white striped jumpsuit, luggage, swimsuit, purple wrap skirt, shorts and top, camera, and posters.

©1988 NRFB $55.00

VACATION SENSATION BARBIE #1675

Barbie doll travels with sportswear and accessories including a blue jumpsuit, luggage, swimsuit, skirt, shorts, camera, and posters.

©1986 NRFB $55.00

Roszella Jones Collection

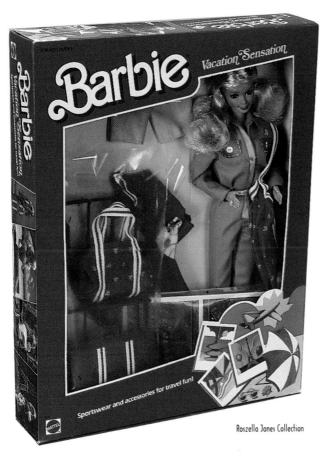

Roszella Jones Collection

Roszella Jones Collection

Clockwise from top left:

WEDDING DAY BARBIE/LOVELY BRIDESMAID #9608
Barbie doll is dressed in a pink bridesmaid dress and
holds a bouquet of pink silk roses.
©1990   NRFB $23.00

WEDDING FANTASY BARBIE   #2125
Barbie doll is dressed in a white and iridescent floor
length gown, with high-neck bodice and large lace
sleeves. Her veil is held to her head with a halo of pearls.
This doll is featured on the cover.
©1989   NRFB $39.00

WESTERN BARBIE   #1757
This doll is dressed in a white jumpsuit trimmed with silver and black and comes with white cowboy boots and
hat. She is a unique doll because she will wink!
©1980   NRFB $30.00

WESTERN FUN BARBIE   #9932
This doll is dressed in turquoise tights, pink fringed jacket, pink, magenta, yellow, and turquoise Western print
skirt, pink felt hat, and pink cowboy boots.
©1989   NRFB $27.00

Roszella Jones Collection

Roszella Jones Collection

Clockwise from top left:

WET 'N WILD BARBIE   #4103
This doll's pink and orange swimsuit will change colors in cold water. She comes with two pieces of jewelry and sunglasses.
©1989   NRFB $18.00

WINTER FANTASY BARBIE (F.A.O. SCHWARZ)   #5946
This doll is the second in a series of F.A.O. Schwarz Limited Edition dolls. She is wearing a wonderful blue satin evening gown trimmed in white fur.
©1989   NRFB $240.00

WINTER FUN BARBIE   #5949
This Toys "Я" Us wintry doll is dressed in white leggings, glittering white shirt and white fur-trimmed jacket, hat, and boots. She comes with skis, ski poles, and sunglasses.
©1990   NRFB $50.00

GIFT SETS

These great sets are usually packaged with either two or more dolls, a doll and an animal, or a doll and another unique extra to make them special gifts. Disney sponsored a gift set that featured Barbie, Ken, and Skipper dolls complete with Mouse-ear caps to wear along with their Disney fashions! Barbie and Ken Go Campin' dolls comes in a set with lots of extras and a My First Barbie doll is offered in a set with 12 extra fashions. One set that is among the most popular is the Midge Wedding Party Gift Set. It includes six great dolls, flower girl Kelly, ring bearer Todd, best man Ken, bridesmaid Barbie, groom Alan, and beautiful bride Midge. I enjoy this particular set as much today as the day I bought it. Perhaps it is because there are so many dolls included and the fashions are impressive. As advertised, it is "all the important people in the year's most glamorous wedding."

On the primary market these sets were a good value because of all of the extras included. Values are for never removed from box (NRFB) items. (See pricing explanation pages 10–11.) To exact the average values reported here, the box as well as the contents must be like new.

Clockwise from top left:

BARBIE & FRIENDS GIFT SET   #4431
This is a set that includes Barbie, P.J., and Ken dolls. P. J. doll is wearing a blue and white summer dress tied at the waist with a pink sash. Ken doll's pink cotton knit shirt is embroidered with his name and his slacks are woven burgundy cotton. Barbie doll's fashion is a blue, green, pink, yellow, and white horizontally striped cotton knit top and her skirt is from the same material as Ken doll's shirt.
1983   NRFB $53.00

BARBIE & KEN CAMPIN' OUT SET   #4984
This is a set that includes Barbie and Ken dolls along with camping items that total more than 20. Barbie doll and Ken doll wear white and red coordinating summer shorts outfits.
1983   NRFB $100.00

DANCE MAGIC BARBIE & KEN GIFT SET   #9058
Both dolls are dressed in white, pink, and iridescent ballroom fashions. Ken doll's hair and Barbie doll's lips change color and extra fashions are included in this set.
©1990   NRFB $48.00

Roxzella Jones Collection

Roxzella Jones Collection

Author's Collection

Clockwise from top left:

DANCE SENSATION BARBIE GIFT SET   #9058
This Toys "Я" Us set comes with several dance outfits in shades of pink, magenta, and purple. You could mix and match the extras for more than 10 fashion looks.
1984   NRFB $38.00

DENIM FUN COOL CITY BLUES GIFT SET   #4893
This is a Limited Edition set from Toys "Я" Us and includes Barbie, Ken, and Skipper dolls dressed in pink and blue denim outfits.
©1989   NRFB $58.00

BARBIE & FRIENDS DISNEY GIFT SET   #3177
This is a Limited Edition set from Toys "Я" Us and includes Barbie, Ken, and Skipper dolls dressed in Disney outfits and Mickey Mouse ears!
©1991   NRFB $70.00

Clockwise from top right:

BARBIE DRESS 'N PLAY   #7543

This amazing gift set includes a winter ski set, exercise center, luggage, clothing, and lots of extras. The box carries no date and is produced by ARCO but the items contained and the Barbie logo suggest it was produced in this '81–'91 decade.

NRFB   No Value Available

FLIGHT TIME BARBIE GIFT SET/HISPANIC   #2066

This set features Barbie doll in a pink flight attendant's outfit with a paper doll modeling a skirt and scarf for the actual doll, a pink briefcase, and pink wings for a child to wear.

©1989   NRFB $40.00

FLIGHT TIME BARBIE GIFT SET/WHITE   #9584

This set features Barbie doll in a pink flight attendant's outfit with a paper doll modeling a skirt and scarf for the actual doll, a pink briefcase, and pink wings for a child to wear.

©1989   NRFB $37.00

Roszella Jones Collection

Roszella Jones Collection

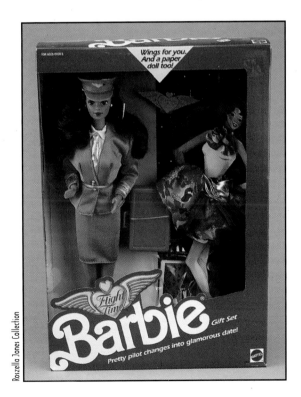

Roszella Jones Collection

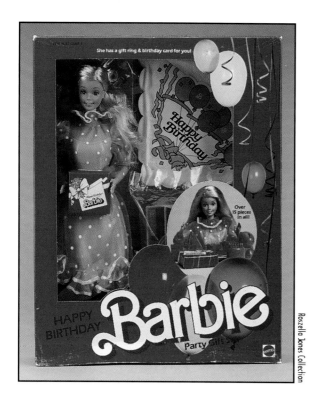

Rozella Jones Collection

Rozella Jones Collection

Rozella Jones Collection

Clockwise from top left:

HAPPY BIRTHDAY BARBIE GIFT SET   #9519
This doll is posed holding a surprise birthday package. Her dress is pink with white flocked polka dots and there are over 15 pieces including pendant, ring, birthday card, cake, plates and cups, tablecloth, and napkins.

1984   NRFB $72.00

I LOVE BARBIE
This Japanese Barbie doll comes in a gift set with cups, saucers, spoons, forks, plates, tea pot, tray, napkins, napkin rings, and a booklet. This set comes in six styles.

©1990   NRFB $65.00

LOVING YOU BARBIE GIFT SET   #7583
This doll features the very recognizable red and white fashion with a red velvet bodice, and white skirt with sheer white overskirt accented with flocked red hearts. Included in this gift set is a child-size purse and heart pendant. Extras include stationery, stamper, and stickers.

©1984   NRFB $60.00

Clockwise from top right:

MY FIRST BARBIE DELUXE FASHION GIFT SET   #2483

This easy-to-dress doll comes with over 12 easy-on fashions. The perfect starter set for little girls, including casual, dress-up, and ballet fashions for hours of fun.

©1991   NRFB $20.00

BARBIE FOR PRESIDENT GIFT SET   #3722

This Toys "Я" Us set features Barbie doll in a red, white, blue, and silver gown and an extra suit of red velvet and a white briefcase. Two versions of this box were issued, the first with the presidential seal and the second with a star. Since Barbie doll was running for president and not actually in office, the seal was removed. The first sometimes commands a higher price.

©1991   NRFB $ 50.00

BARBIE SHARIN' SISTERS GIFT SET   #5716

This gift set includes Stacie, Barbie, and Skipper dolls and fashions that they could share. You could combine the clothing for over 30 fun fashion looks.

©1991   NRFB $37.00

Roszella Jones Collection

Roszella Jones Collection

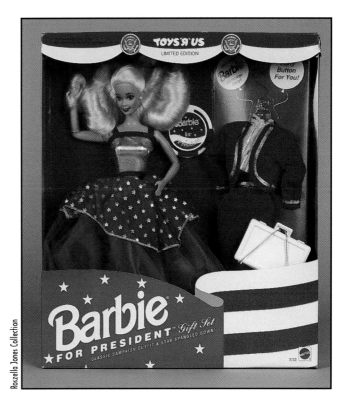

Roszella Jones Collection

Stars 'n Stripes

MARINE CORPS
Barbie & Ken

Rozella Jones Collection

Tennis Stars
Barbie & Ken
Doubles, singles...they're the champs!

MATTEL

Rozella Jones Collection

SPECIAL!
Includes lovely,
long gown and sparkly
jumpsuit!

Twirly Curls
Barbie Gift Set

Style her hair in fabulous twists 'n twirls!

MATTEL

Rozella Jones Collection

Clockwise from top left:

STARS 'N STRIPES MARINE CORPS BARBIE & KEN SET   #4704
In this Stars 'n Stripes Special Edition set, Barbie and Ken dolls are in dress uniforms each with a multi-stripe Desert Storm medal for participation in that campaign!

1991	NRFB	$ 57.00

TENNIS STARS BARBIE & KEN   #7801
This package includes 2 dolls, tennis clothing, tennis shoes, net and stands, rackets and more.

©1986	NRFB	$55.00

TWIRLY CURLS BARBIE GIFT SET   #4097
This gift set features Barbie doll dressed in a hot pink layered gown. Included in the package are a glittery pink jumpsuit, Twirly Curler hairstyler, chair, comb, brush, 4 barrettes, and 2 ribbons.

©1982	NRFB	$60.00

All the important people in the year's most glamorous wedding!

Wedding Party
Midge
GIFT SET

MIDGE WEDDING PARTY GIFT SET #9852
This gift set comes with the whole wedding party! Flower girl Kelly, ring bearer Todd, best man Ken, bridesmaid Barbie, groom Alan and Midge as the beautiful bride! Six great dolls in beautifully styled fashions for a special day.
1990 NRFB $125.00

FASHIONS

Fashions packaged without the doll are nostalgic reminders of the first marketing strategy of Barbie doll. The idea was to sell the doll and then sell the fashions separately, so these fashions are especially intriguing. There were several Collector Series and the flashy Rocker fashions were real showstoppers. The space race was reflected in the Astro Series and even pets entered the vogue scene in Pet Show Fashions. Oscar de la Renta designed a series of exquisite evening gowns. From City Nights to Western Wear, these fashions are captivating and unusual collectibles.

It is very tempting and often desirable from a visual point of view to remove these fashions and display them on a doll. But as with all of these collecting fields, values are for never removed from box (NRFB) items. (See pricing explanation pages 10–11.) To exact the average values reported here, the box as well as the contents must be like new.

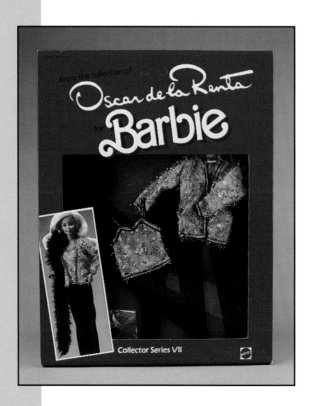

Author's Collection

Counterclockwise from top:

3 FABULOUS BALLGOWNS   #3714   ©1991   Mint $30.00
Offered in the J. C. Penney catalog, these three evening fashions were sold as a set without dolls and had to be ordered through the catalog.

BARBIE ASTRO FASHIONS — DAZZLING DANCER   #2743   ©1985   NRFP $28.00
Metallic midnight blue and silver gown, stockings, and knee-high boots complete this Astro fashion package.

BARBIE ASTRO FASHIONS — GALAXY A GO GO   #2742   ©1985   NRFP $28.00
Silver and white minidress, silver and pink overcoat, white and silver over-the-knee boots, and fashion accessories complete this Astro fashion package.

Roszella Jones Collection

Roszella Jones Collection

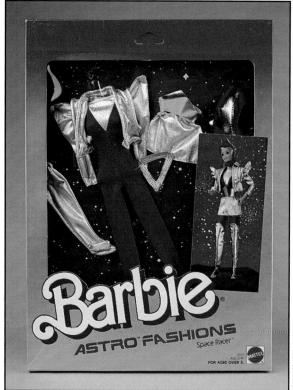

Clockwise from top left:

BARBIE ASTRO FASHIONS — SPACE RACER   **#2737**
Silver and red jumpsuit, silver miniskirt, red and silver hat, silver over-the-knee boots, and fashion accessories complete this Astro fashion package.
©1985   NRFP $28.00

BARBIE ASTRO FASHIONS — STARLIGHT SLUMBERS   **#2739**
Silver and white floor-length dress with white pleated over-dress, and pink open-toe high heels complete this Astro fashion package.
©1985   NRFP $28.00

BARBIE ASTRO FASHIONS — WELCOME TO VENUS   **#2738**
Silver and metallic pink outfit with pink leggings, pink over-the-knee boots, and fashion accessories complete this Astro fashion package.
©1985   NRFP $28.00

BARBIE BATH FUN FASHION PLAYSET #9266
Pink robe, towels, towel rack, hangers, scale, sponge, make-up bag, shampoo bottle, and more accessories complete this Barbie Bath fashion package.
©1984   NRFP $15.00

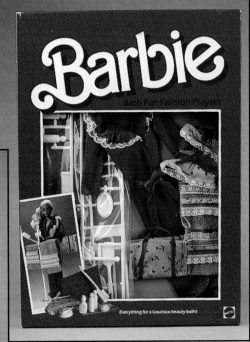

Roszella Jones Collection

Author's Collection

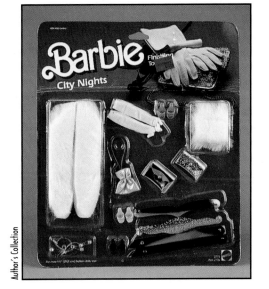

Author's Collection

BARBIE CHEERLEADER SET   #7278   ©1990   NRFP $8.00
This is a card of cheerleading accessories including a drum, baton, hat, and more.

BARBIE CITY NIGHTS FINISHING TOUCHES   #2773   ©1985   NRFP $15.00
This is a card of fashion accessories including a fur wrap, muff, shoes, jewelry, belts, gloves, and purses.

Rozella Jones Collection

Rozella Jones Collection

Clockwise from top left:

BARBIE COLLECTOR SERIES I — HEAVENLY HOLIDAYS   #4277
Red velvet and silver trimmed cape and long skirt, white lace under-dress, plaid satin cummerbund, white open-toe high heels, wrapped gift, and more accessories complete this Barbie fashion package.
©1982   NRFP $55.00

BARBIE COLLECTOR SERIES II — SPRINGTIME MAGIC   #7092
Strapless gown of white, purple, pink, and silver, ruffle boa, straw hat, basket of flowers, pink open-toe high heels and fashion accessories make up this Collector fashion package. This fashion was sold on the doll in Germany. (See Foreign Fruhlingszauber Barbie doll in the Doll section.)
©1983   NRFP $50.00

BARBIE COLLECTOR SERIES III — SILVER SENSATION   #7438
Silver strapless floor-length straight dress with a gathered magenta floor-length overskirt, magenta cape, pink open-toe high heels, silver gloves, and pink Barbie hanger make up this Collector fashion package.
©1983   NRFP $40.00

Clockwise from top right:

BARBIE DAY-TO-NIGHT FASHIONS — BUSINESS EXECUTIVE   #9083
Teal and dark magenta suit dress, pink satin evening gown, purse, and pink heels complete this Barbie fashion package.
©1984   NRFP $15.00

BARBIE DAY-TO-NIGHT FASHIONS — DANCER   #9082
White ribbed gown, magenta overskirt, magenta jumpsuit, and white heels complete this Barbie fashion package.
©1984   NRFP $15.00

BARBIE DAY-TO-NIGHT FASHIONS — DRESS DESIGNER   #9081
Black skirt, white, red, and black jacket, white felt hat, black portfolio, red evening skirt, red heels, and accessories complete this Barbie fashion package.
©1984   NRFP $15.00

Roszella Jones Collection

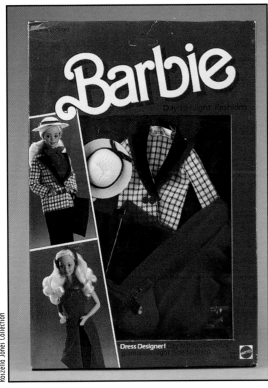

Roszella Jones Collection

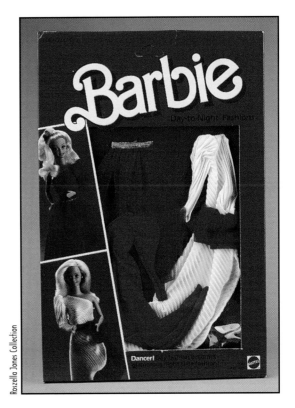

Roszella Jones Collection

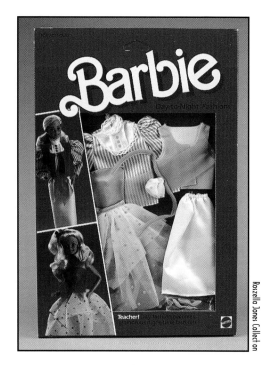

Roszella Jones Collection

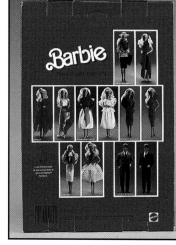

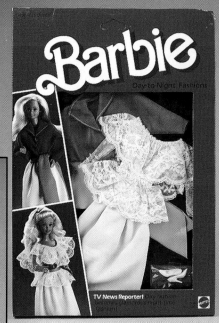

Roszella Jones Collection

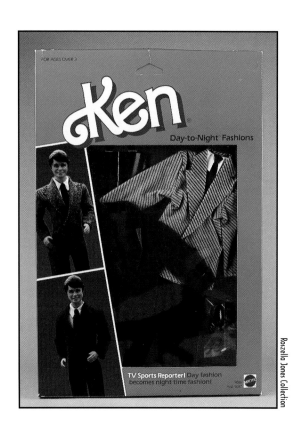

Roszella Jones Collection

Clockwise from top left:

BARBIE DAY-TO-NIGHT FASHIONS — TEACHER　　　#9085
Lavender and white blouse, white skirt, lavender vest, white and purple sheer overskirt, lavender heels, and accessories complete this Barbie fashion package.
©1984　　　　　　NRFP　　　　　　$15.00

BARBIE DAY-TO-NIGHT FASHIONS — TV NEWS REPORTER　　#9084
White lacy blouse, red satin blouse, white skirt, white heels, and accessories complete this Barbie fashion package.
©1984　　　　　　NRFP　　　　　　$15.00

KEN DAY-TO-NIGHT FASHIONS — TV SPORTS REPORTER　　#9086
Burgundy velvet jacket, black pin stripe jacket, black slacks, white shirt, burgundy tie, daytimer, and black shoes complete this Ken fashion package.
©1984　　　　　　NRFP　　　　　　$15.00

Roszella Jones Collection

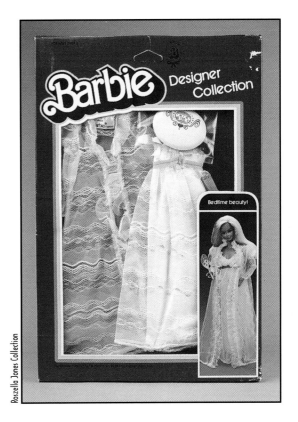

Roszella Jones Collection

Clockwise from top left:

BARBIE DESIGNER COLLECTION — AFTERNOON PARTY   #5835
A white dress trimmed in pink plaid with white lace ruffles and leg-of-mutton sleeves, underskirt with pink plaid trim, and pink cowboy boots make up this fashion package.
©1982   NRFP $18.00

BARBIE DESIGNER COLLECTION — BEDTIME BEAUTY   #7081
White gown, sheer white robe, hairbrush, scale, open-toe white shoes, and pink Barbie hanger make up this fashion package.
©1983   NRFP $10.00

BARBIE DESIGNER COLLECTION — DATE NIGHT   #5654
A shimmering party dress with white and pink overlay and wrap, and pink open-toe shoes make up this fashion package.
©1982   NRFP $18.00

Roszella Jones Collection

Clockwise from top left:

BARBIE DESIGNER COLLECTION — HORSEBACK RIDING   **#7080**
White riding pants, pink shirt, pink striped navy jacket, navy velvet skirt, navy boots, navy hat, riding crop, and pink Barbie hanger make up this fashion package.
©1983   NRFP $10.00

BARBIE DESIGNER COLLECTION — IN THE SPOTLIGHT   **#7082**
A white formal jumpsuit with iridescent cummerbund and pink velvet bow tie, white jacket with iridescent collar, fur wrap, iridescent hat, white heels, and pink Barbie hanger make up this fashion package.
©1983   NRFP $10.00

BARBIE DESIGNER COLLECTION — PICTURE IN PLAID   **#7083**
Teal plaid satin gown with mauve ribbed strapless bodice, teal plaid wrap, mauve heels and white Barbie hanger make up this fashion package.
©1983   NRFP $10.00

Rozella Jones Collection

Roszella Jones Collection

Roszella Jones Collection

Clockwise from top left:

BARBIE DESIGNER COLLECTION — SKI PARTY   #7079
Gray knit jumpsuit, gray fur trimmed hood, pink vinyl fur trimmed jacket, mirrored sunglasses, pink vinyl bag, skis and ski poles, pink ski boots, and white Barbie hanger make up this fashion package.
©1983   NRFP $10.00

KEN DESIGNER COLLECTION — SIMPLY DASHING   #7084
Jumpsuit of black slacks and white top with pink tie, black jacket with pink boutonniere, black socks and shoes make up this fashion package.
©1983   NRFP $10.00

BARBIE DESIGNER ORIGINALS — FUN 'N FANCY   #3800
A pale yellow underdress with ruffle, blue velvet overskirt with red waist sash, and pale yellow open-toe shoes make up this fashion package.
©1981   NRFP $20.00

Roszella Jones Collection

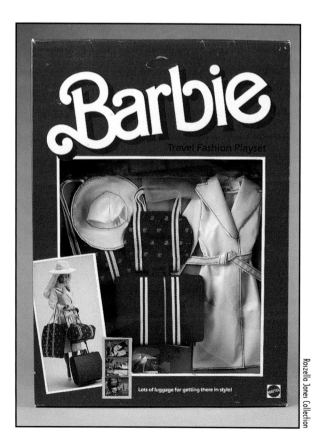

Clockwise from top left:

BARBIE DESIGNER ORIGINALS — WHITE DELIGHT   #3799
A white jumpsuit with pale lavender belt, a white long jacket with silver polka dots and lavender trim, and white open-toe shoes make up this fashion package.
©1981   NRFP $20.00

KEN DESIGNER ORIGINALS — DANDY LINES   #3797
Pin stripe jacket and slacks, burgundy tie and white shirt front, and black shoes make up this fashion package.
©1981   NRFP $20.00

BARBIE TRAVEL FASHION PLAYSET   #9264
White vinyl raincoat and hat, suitcase and garment bag, duffle bag, luggage carrier, camera, scarf, white heels, and more accessories complete this Barbie Travel fashion package.
©1984   NRFP $15.00

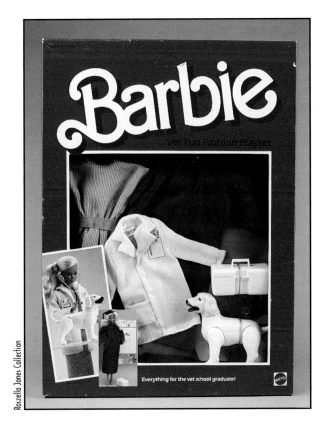

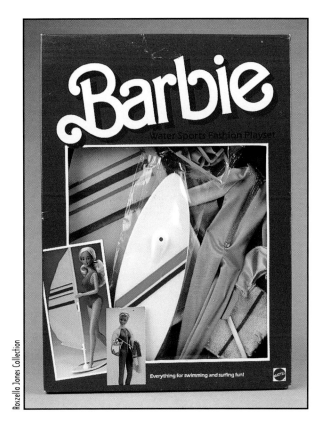

Clockwise from top left:

BARBIE VET FUN FASHION PLAYSET   #9267
Graduation robe and hat, white lab jacket, pink and white striped dress, stethoscope, shoes, dog, dog show certificate, and more accessories complete this Barbie Vet fashion package.
©1984   NRFP $15.00

BARBIE WATER SPORTS FASHION PLAYSET   #9263
Wet suit with hood, pink swimsuit, surfboard and sail, towel, swim fins and goggles, snorkel, face mask, sunglasses, and more accessories complete this Barbie Sports fashion package.
©1984   NRFP $15.00

KEN THE JEANS LOOK FASHIONS — MODE JEAN   #4336
Blue jeans and blue jean jacket, white T-shirt with pink, blue, and metallic print, and brown boots complete this Ken fashion package.
©1987   NRFP $12.00

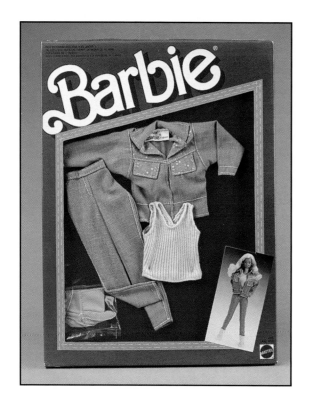

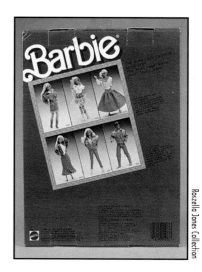

Top to bottom:

**BARBIE THE JEANS LOOK FASHIONS —
MODE JEAN** #4335
Blue jeans and blue jean jacket, pink tank top, and
pink boots complete this Barbie fashion package.
©1987 NRFP $12.00

**FROM THE COLLECTION OF OSCAR DE LA RENTA
FOR BARBIE** #9258
Red satin gown with gold metallic accents, red ruf-
fle-neck jacket, and red heels complete this Collector Series IV fashion package.
©1984 NRFP $35.00

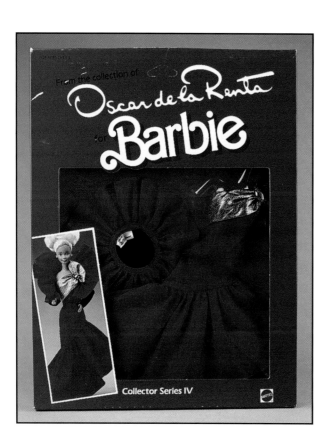

Collector Series V

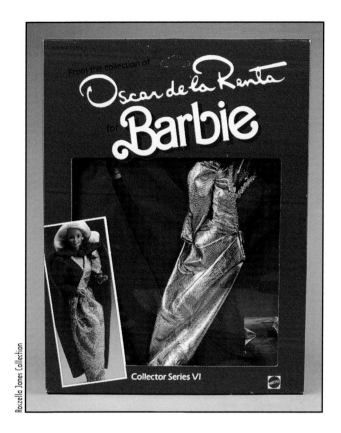

Collector Series VI

Clockwise from top left:

FROM THE COLLECTION OF OSCAR DE LA RENTA FOR BARBIE #9259
Blue, teal, and green evening gown with a metallic sheen, teal satin cape, purse, and heels complete this Collector Series V fashion package.
©1984   NRFP $35.00

FROM THE COLLECTION OF OSCAR DE LA RENTA FOR BARBIE #9260
Purple satin cape, metallic gold and purple satin evening gown, and purple heels complete this Collector Series VI fashion package.
©1984   NRFP $35.00

FROM THE COLLECTION OF OSCAR DE LA RENTA FOR BARBIE #9261
Black velvet floor-length straight skirt, black stockings, yellow and metallic gold top and jacket, black fur boa, and black heels complete this Collector Series VII fashion package.
©1984   NRFP $35.00

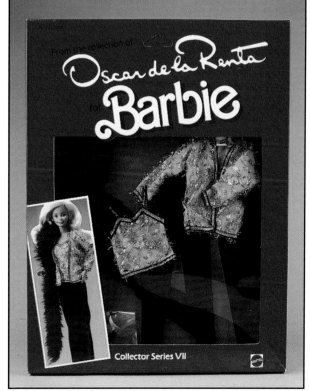

Collector Series VII

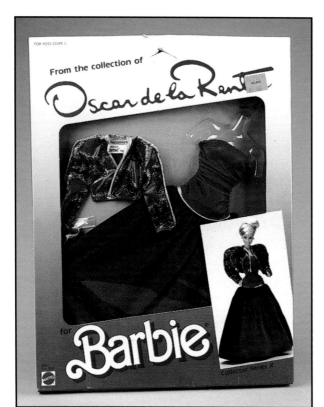

Clockwise from top left:

FROM THE COLLECTION OF OSCAR DE LA RENTA FOR BARBIE   #2762
Blue strapless evening gown accented in silver, blue fur boa, and blue heels complete this Collector Series VIII fashion package.
©1985   NRFP $30.00

FROM THE COLLECTION OF OSCAR DE LA RENTA FOR BARBIE   #2763
A burgundy and gold metallic gown, red satin and gold metallic jacket, gold belt, and red heels complete this Collector Series IX fashion package.
©1985   NRFP $30.00

FROM THE COLLECTION OF OSCAR DE LA RENTA FOR BARBIE   #2765
Deep purple velvet and magenta satin strapless gown, flashy magenta, purple, and metallic gold jacket, and purple heels complete this Collector Series X fashion package.
©1985   NRFP $30.00

Clockwise from top right:

FROM THE COLLECTION OF OSCAR DE LA RENTA FOR BARBIE   #2766
Two choices come with this Collector Series XI fashion package. One is a street-length magenta lace and satin dress accented with metallic gold. The other is a floor-length magenta lace and satin gown with a metallic sash at the hip. Magenta heels complete the outfits.
©1985   NRFP $30.00

FROM THE COLLECTION OF OSCAR DE LA RENTA FOR BARBIE   #2767
Midnight blue, and metallic blue with gold accents make this Collector Series XII fashion package special. Dark blue heels complete the outfit.
©1985   NRFP $30.00

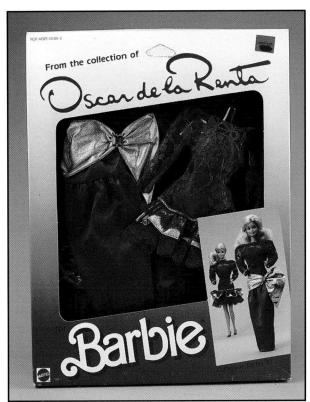

Roszella Jones Collection

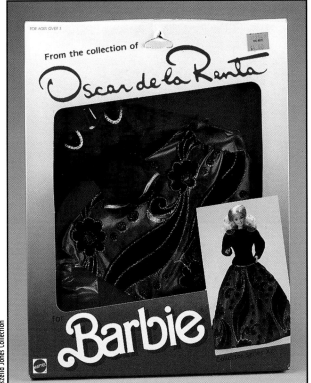

Roszella Jones Collection

Clockwise from top left:

BARBIE PET SHOW FASHIONS #3657
Iridescent pastel green top and white iridescent pleated skirt, white shoes, and soft white kitten complete this Pet Show fashion package.
©1986 NRFP $15.00

BARBIE PET SHOW FASHIONS #3658
Pastel yellow and metallic gold top, yellow skirt, yellow shoes, and soft white kitten complete this Pet Show fashion package.
©1986 NRFP $15.00

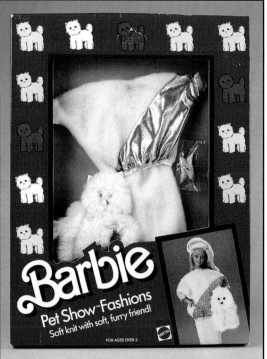

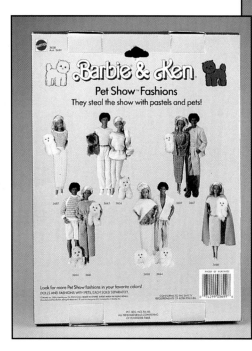

Clockwise from top left:

KEN PET SHOW FASHIONS #3664
Pastel yellow sweater vest, yellow slacks, white shoes, and soft white kitten complete this Pet Show fashion package.
©1986 NRFP $15.00

KEN PET SHOW FASHIONS #3667
Pastel pink shirt, blue linen jacket and slacks, white shoes, and soft gray kitten complete this Pet Show fashion package.
©1986 NRFP $15.00

BARBIE PRIVATE COLLECTION FASHIONS #4509
White fur coat, metallic gold stockings, gold purse, white heels, and more accessories complete this Barbie fashion package.
©1988 NRFP $28.00

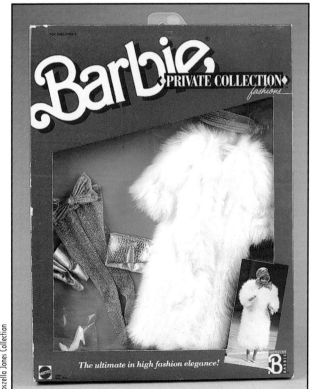

Author's Collection

Author's Collection

Rozella Jones Collection

Clockwise from top left:

BARBIE PRIVATE COLLECTION FASHIONS   #4961
This is a gold lamé coat with white fur trim and sleeves, gold lamé purse, and hat.
©1989   NRFP $28.00

BARBIE PRIVATE COLLECTION FASHIONS   #7096
This fashion has a magenta fur trimmed teal wrap, teal skirt, a blouse made of black, purple, magenta, and white print, and a magenta felt hat.
©1990   NRFP $28.00

BARBIE PRIVATE COLLECTION FASHIONS   #7097
Metallic riot-of-color full skirt, top with dark metallic bodice, puffed sleeves, and magenta velvet trim, magenta heels, and more accessories complete this Barbie fashion package.
©1990   NRFP $28.00

Author's Collection

Clockwise from top right:

BARBIE PRIVATE COLLECTION FASHIONS   **#7113**
This is a red and purple evening fashion with a short coat of metallic gold, purple, red, and black. It comes with a scarf and gold purse.
©1990   NRFP $28.00

BARBIE AND THE ROCKERS FASHIONS   **#1165**
Hot pink minidress, sparkling purple jacket, lime scarf, purple over-the-knee leather boots and fashion accessories make up this Rocker fashion package.
©1985   NRFP $18.00

BARBIE AND THE ROCKERS FASHIONS   **#1166**
Hot pink leggings, hot pink tank top, pink and purple jacket, yellow tube top, purple heels and fashion accessories make up this Rocker fashion package.
©1985   NRFP $18.00

Roszella Jones Collection

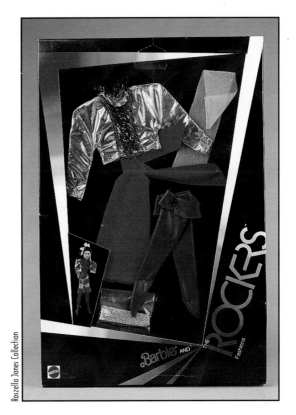

Roszella Jones Collection

Clockwise from top left:

BARBIE AND THE ROCKERS FASHIONS #1167

Pink stockings, blue miniskirt, white and pink knit top, pink jacket, blue felt hat, blue heels, and fashion accessories make up this Rocker fashion package.

©1985   NRFP $18.00

BARBIE AND THE ROCKERS FASHIONS #1170

Blue sweatshirt, yellow miniskirt, white and pink leggings, yellow scarf, and lime heels make up this Rocker fashion package.

©1985   NRFP $18.00

BARBIE AND THE ROCKERS FASHIONS #1175

Pink leggings, pink tube top, pink and silver shirt, print tie top, yellow scarf, pink heels and fashion accessories make up this Rocker fashion package.

©1985   NRFP $18.00

Rozella Jones Collection

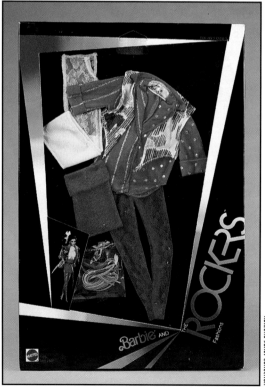

Rozella Jones Collection

Rozella Jones Collection

Clockwise from top left:

BARBIE AND THE ROCKERS FASHIONS #1177
Color-splashed white slacks, black shirt, lime leather jacket, orange socks and cummerbund, and white tennis shoes make up this Ken Rocker fashion package.
©1985 NRFP $18.00

BARBIE AND THE ROCKERS FASHIONS #1176
Orange leggings, black leather miniskirt, hot pink top, white color-splashed jacket, and pink heels make up this Rocker fashion package.
©1985 NRFP $18.00

BARBIE & THE ROCKERS FASHIONS #2688
White party dress with sheer overlay accented with sparkling blue, pink, and green dots, white veil, white heels, and fashion accessories make up this Rocker fashion package.
©1985 NRFP $18.00

Roszella Jones Collection

Roxella Jones Collection

Roxella Jones Collection

Clockwise from top left:

BARBIE AND THE ROCKERS FASHIONS   #2689   ©1985   NRFP $18.00
Yellow sheer leggings, hot pink ruffled skirt, yellow off-the-shoulder top, pink heels, and fashion accessories make up this Rocker fashion package.

BARBIE AND THE ROCKERS FASHIONS   #2690   ©1985   NRFP $18.00
Lime lace leggings, hot pink minidress, blue metallic jacket, pink heels, and fashion accessories make up this Rocker fashion package.

BARBIE AND THE ROCKERS FASHIONS   #2691   ©1985   NRFP $18.00
White slacks, pink shirt front with blue metallic cummerbund, blue satin jacket, and white tennis shoes make up this Ken Rocker fashion package.

BARBIE AND THE ROCKERS FASHIONS   #2791   ©1985   NRFP $18.00
Color-splashed white slacks, yellow shirt, orange tube top, yellow tennis shoes, pink socks, and fashion accessories make up this Rocker fashion package.

Roxella Jones Collection

Roxella Jones Collection

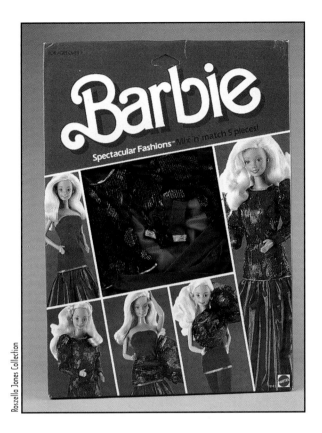

Clockwise from top left:

BARBIE SPECTACULAR FASHIONS　　　　　#9143
Metallic gold, midnight blue, and purple satin tube top, pants, blouse, 2 skirts and shoes make up this mix-and-match 5-piece fashion package.
©1984　　　　　NRFP　　　　　$12.00

BARBIE SPECTACULAR FASHIONS　　　　　#9144
Metallic multicolor, light teal, leather-and-fur-trimmed jacket, hood, minidress, belt, wrap skirt, long skirt, and shoes make up this mix-and-match 7-piece fashion package.
©1984　　　　　NRFP　　　　　$12.00

BARBIE SPECTACULAR FASHIONS　　　　　#9145
Red, white, and metallic gold tube top, pleated skirt, blouse, harem pants, top, overskirt, gloves, leggings, and shoes make up this mix-and-match 8-piece fashion package.
©1984　　　　　NRFP　　　　　$12.00

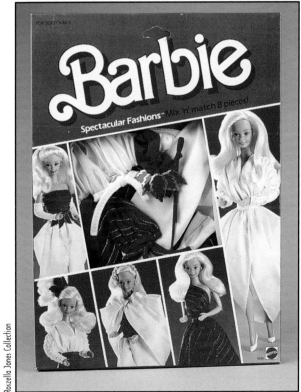

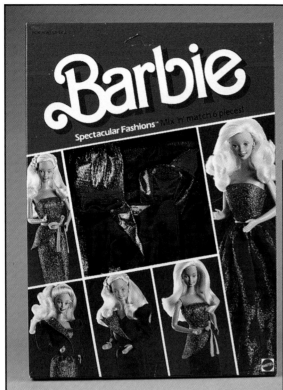

Roszella Jones Collection

BARBIE SPECTACULAR FASHIONS   #9146
Metallic pink and burgundy satin jacket, top, long skirt, short skirt, overskirt, peplum and shoes make up this mix-and-match 6-piece fashion package.
©1984   NRFP $12.00

BARBIE SPECTACULAR FASHIONS — BLUE MAGIC   #7216
Blue slacks, blue top with iridescent belt, white fur muff, white sheer fur trimmed tutu, white sheer fur trimmed gathered long skirt, white ice skates, blue open-toe shoes, and white Barbie hanger make up this mix-and-match fashion package.
©1983   NRFP $12.00

Roszella Jones Collection

Clockwise from top left:

BARBIE SPECTACULAR FASHIONS — DANCE SENSATION   #7218
Magenta, pink, and purple mix-and-match pieces, magenta ballerina shoes, and purple open-toe shoes make up this fashion package.
©1983   NRFP $12.00

BARBIE SPECTACULAR FASHIONS — IN THE PINK   #7219
Hot pink long, straight skirt, metallic pink trimmed white jacket, pink and silver bathing suit, tennis bag and racket, pink tennis ball, pink ruffle boa, sheer pink overskirt, white tennis shoes, and pink open-toe shoes make up this mix-and-match fashion package.
©1983   NRFP $12.00

BARBIE SPECTACULAR FASHIONS — RED SIZZLE   #7217
Sparkling red form-fitting gown, long gathered red and silver skirt, silver blouse, silver ruffle, red and silver fashion accessories, fur boa, and 2 pairs of open-toe shoes make up this mix-and-match fashion package.
©1983   NRFP $12.00

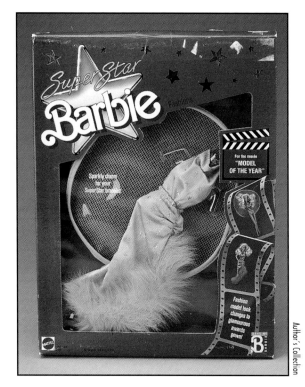

Author's Collection

SUPERSTAR BARBIE MODEL OF THE YEAR #3301
This is the lavender SuperStar fashion with lavender fur trim and exaggerated net headwear.
©1988   NRFP $12.00

Roszella Jones Collection

BARBIE WEDDING OF THE YEAR — BRIDESMAID'S DREAM #5745
Pastel teal floor-length dress with ruffle collar, bouquet, white open-toe high heels, and accessories complete this Barbie fashion package.
©1982   NRFP $18.00

Roszella Jones Collection

BARBIE ROMANTIC FASHIONS #3102
Long, white wedding dress and hat with veil, pink rose bouquet, white heels, and fashion accessories make up this wedding fashion package. Featured on the back is a photograph of the other fashions in this collection.
©1986   NRFP $20.00

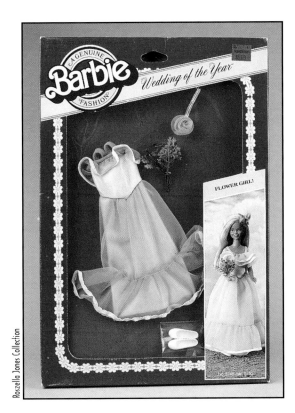

Roszella Jones Collection

BARBIE WEDDING OF THE YEAR — FLOWER GIRL　　　**#5746**
Pastel pink floor-length dress, bouquet, white shoes, and accessories complete this Skipper fashion package.
©1982　　　　　　　　　NRFP　　　　　　$18.00

Roszella Jones Collection

BARBIE WEDDING OF THE YEAR — SUITED FOR THE GROOM　　**#5744**
White tuxedo with iridescent collar, white slacks, white shirt front with ruffle, white cummerbund, white shoes, and accessories complete this Ken fashion package. Featured on the back of the package is a photo of the other fashions in this collection. Also featured is the Designer Collection.
©1982　　　　　　　　　NRFP　　　　　　$14.00

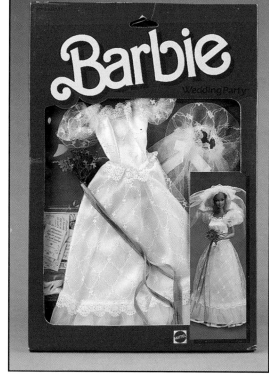

Clockwise from top left:

BARBIE WEDDING OF THE YEAR — HERE COMES THE BRIDE   #5743
White floor-length wedding dress, white veil, bouquet, white open-toe high heels, and accessories complete this Barbie fashion package.
©1982   NRFP $20.00

BARBIE WEDDING PARTY   #7965
White floor-length wedding dress, white veil, bouquet, invitations, white heels, and fashion accessories complete this Barbie fashion package.
©1984   NRFP $12.00

BARBIE WEDDING PARTY   #7966
White jacket, black slacks, white shirt front, pink cummerbund, black socks and shoes, and accessories complete this Ken fashion package.
©1984   NRFP $10.00

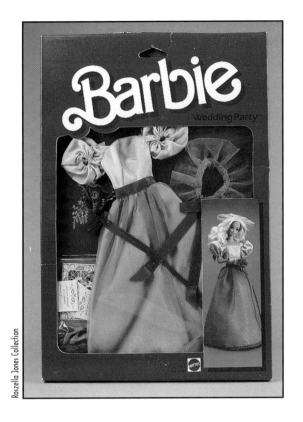

Roszella Jones Collection

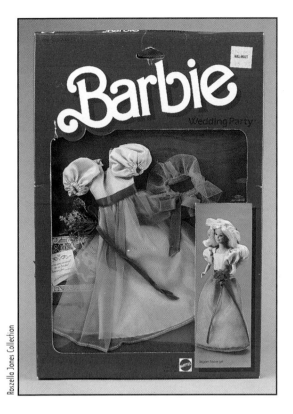

Roszella Jones Collection

Clockwise from top left:

BARBIE WEDDING PARTY   #7967
Pink floor-length bridesmaid's dress, pink net hat, bouquet, invitations, pink heels, and fashion accessories complete this Barbie fashion package.
©1984   NRFP $12.00

BARBIE WEDDING PARTY — SKIPPER FLOWER GIRL   #7968
Pink floor-length flower girl dress, bouquet, invitation, and fashion accessories complete this Skipper fashion package.
©1984   NRFP $10.00

BARBIE & KEN WESTERN FASHIONS   #3578
Navy and rose calico skirt and top, brown leather vest and purse, brown hat, and brown cowboy boots make up this Barbie fashion package.
©1981   NRFP $9.00

Roszella Jones Collection

Clockwise from top left:

BARBIE & KEN WESTERN FASHIONS #3578
Denim blue jeans, magenta cowboy shirt, silver belt, white hat, and white cowboy boots make up this Barbie fashion package.
©1981   NRFP $9.00

BARBIE & KEN WESTERN FASHIONS #3579
Burgundy leather pants and jacket, white scarf, white hat, and white cowboy boots make up this Barbie fashion package.
©1981   NRFP $8.00

BARBIE & KEN WESTERN FASHIONS #3580
White jeans, burgundy leather-like jacket, white scarf, white hat, and brown cowboy boots make up this Ken fashion package.
©1981   NRFP $8.00

PAPER DOLLS

What a special category since these two-dimensional figures represent the genre that sparked the idea for the Barbie doll in the first place! These cardboard models are most often found in book form with the doll featured on a cardboard page with the clothing on the interior paper pages. Often there were envelopes, cardboard punch out items, or mini posters on the back cover. Some of these collectibles came in a box with the paper doll precut and a plastic stand to hold her. The Peck-Gandré nostalgic line came in a bristol weight envelope with a picture frame chip window on the front. Here you could see the paper doll in this special packaging and artwork of the fashions within from either side of the envelope.

The great thing about collecting paper dolls from this time period is that they are not hard to find. Publishers produced these playthings in tremendous quantities and they were inexpensive. The fashions that are represented give you a good history of the fashions produced for Barbie throughout this decade.

As a general rule in collecting paper dolls, sets that are cut are worth half or less. But values given here are for uncut, mint or never removed from box (NRFB) items. (See pricing explanation pages 10–11.) To exact the average values reported here, the book, envelope or box, as well as the contents, must be like new.

Rozella Jones Collection

Rozella Jones Collection

Rozella Jones Collection

Author's Collection

Clockwise from top left:

ANGEL FACE BARBIE PAPER DOLL #1982-45
This Angel Face Barbie paper doll book includes one 9¾" tall paper doll of a blonde Barbie doll, and fashions. It originally sold for $1.29. Published by Western Publishing Co. A Golden Barbie Paper Doll.
©1983　　　　　　Uncut　　　　　　$5.00

ANGEL FACE BARBIE PAPER DOLL #7407-2E
This Golden boxed paper doll includes one paper doll with a plastic stand, and 15 piece wardrobe. These fashions were precut. Published by Western Publishing Co. A Golden Barbie Paper Doll.
©1983　　　　　　Uncut　　　　　　$9.00

BARBIE PAPER DOLL #1690
This deluxe paper doll book includes one 9¾" tall paper doll of a blonde Barbie doll and fashions. It originally sold in the $2.00 range. Published by Western Publishing Co. A Golden Barbie Paper Doll.
©1990　　　　　　Uncut　　　　　　$3.00

BARBIE PAPER DOLL #1523-2
This paper doll book includes one 9¾" tall paper doll of a blonde Barbie doll, fashions, and press-out box and folder. It originally sold in the $2.00 range. Published by Western Publishing Co. A Golden Barbie Paper Doll.
©1990　　　　　　Uncut　　　　　　$3.00

Clockwise from top right:

BARBIE PAPER DOLL   **#1695**

This fashion paper doll book includes one 9¾" tall paper doll of a blonde Barbie doll and fashions. It originally sold in the $2.00 range. Published by Western Publishing Co. A Golden Barbie Paper Doll.

©1991   Uncut $3.00

BARBIE & KEN PAPER DOLL   **#1985-51**

This Barbie & Ken paper doll book includes one 9¾" tall paper doll of a blonde Barbie doll, Ken doll, and fashions. It originally sold for $1.29. Published by Western Publishing Co. A Golden Barbie Paper Doll.

©1984   Uncut $5.00

CHRISTMAS TIME BARBIE PAPER DOLL   **#1731**

This Christmas Time Barbie paper doll book includes one 9¾" tall paper doll of a blonde Barbie doll and fashions. It originally sold for $1.29. Published by Western Publishing Co. A Golden Barbie Paper Doll. This paper doll may be more in demand for this time period because of the popularity of the Holiday line of Barbie dolls.

©1984   Uncut $16.00

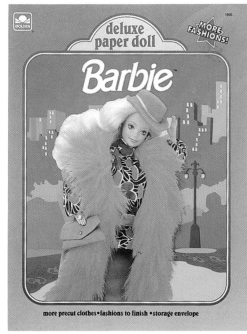

Roszella Jones Collection

Roszella Jones Collection

Roszella Jones Collection

Counterclockwise from top left:

CRYSTAL BARBIE PAPER DOLL   #1983–46
This Crystal Barbie paper doll book includes one 9¾" tall paper doll of a blonde Barbie doll, and fashions. It originally sold for $1.29. Published by Western Publishing Co. A Golden Book Barbie Paper Doll.
©1984   Uncut $5.00

CRYSTAL BARBIE PAPER DOLL   #7407–2B
This Golden boxed paper doll includes one paper doll with a plastic stand, and 14-piece wardrobe. These fashions were precut. Published by Western Publishing Co.
©1984   Uncut $8.00

DAY-TO-NIGHT BARBIE PAPER DOLL   #1982–48
This Day-To-Night Barbie paper doll book includes one 9¾" tall paper doll of a blonde Barbie doll and fashions. It originally sold for $1.29. Published by Western Publishing Co. A Golden Book Barbie Paper Doll.
©1985   Uncut $5.00

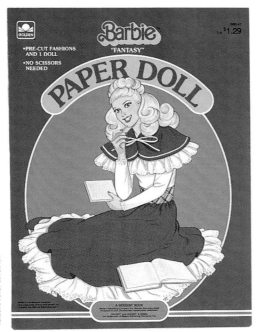

Clockwise from top left:

FANTASY BARBIE PAPER DOLL #1982-47

This Fantasy Barbie paper doll book includes one 9¾" tall paper doll of a blonde Barbie doll and fashions. It originally sold for $1.29. Published by Western Publishing Co. A Golden Book Barbie Paper Doll.

©1984   Uncut $5.00

FANTASY FASHION PAPER DOLL #1502-1

This Fantasy Fashion paper doll book includes one 9¾" tall paper doll of a blonde Barbie doll and fashions. It originally sold in the $2.00 range. Published by Western Publishing Co. A Golden Book Barbie Paper Doll.

©1990   Uncut $4.00

GOLDEN DREAM BARBIE PAPER DOLL #1983-43

This Golden Dream Barbie paper doll book includes one 9¾" tall paper doll of a blonde Barbie doll, and fashions. It originally sold for $1.29. Published by Western Publishing Co. A Whitman Book Barbie Paper Doll.

©1982   Uncut $6.00

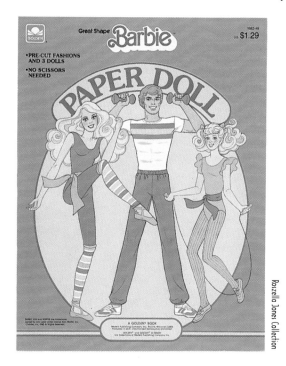

Rozella Jones Collection

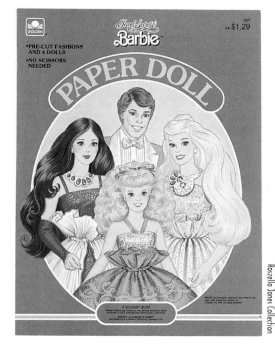

Rozella Jones Collection

Clockwise from top left:

GREAT SHAPE BARBIE PAPER DOLL   #1982-49   ©1986   Uncut $5.00
This Great Shape Barbie paper doll book includes one 9¾" tall paper doll of a blonde Barbie doll, Ken and Skipper dolls, and fashions. It originally sold for $1.29. Published by Western Publishing Co. A Golden Book Barbie Paper Doll.

JEWEL SECRETS PAPER DOLL   #1537   ©1987   Uncut $8.00
This Jewel Secrets paper doll book includes one 9¾" tall paper doll of a blonde Barbie doll, Ken, Skipper, and Whitney dolls, and fashions. It originally sold for $1.39. Published by Western Publishing Co. A Golden Book Barbie Paper Doll.

PECK-GANDRÉ PRESENTS NOSTALGIC BARBIE PAPER DOLLS   ©1989   Uncut $10.00
One 12" paper doll of a brunette ponytail Barbie in her gold and white bathing suit and 16 authentic fashions are enclosed with the Story of Barbie on the back of the package. One 12" paper doll of a blonde ponytail Barbie doll in her black and white bathing suit and 16 authentic fashions are enclosed with the Story of Barbie on the back of the package. One 12" paper doll of a brunette Ken doll in his red bathing suit and fashions are enclosed with the Story of Ken on the back of the package.

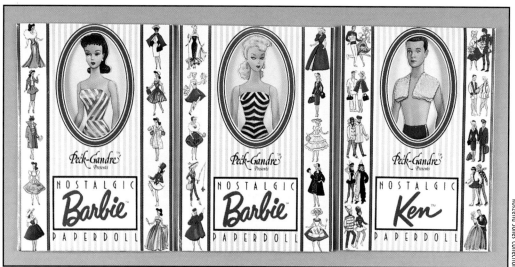

Rozella Jones Collection

Clockwise from top right:

PEACHES 'N CREAM BARBIE PAPER DOLL   #1983-48

This Peaches 'n Cream Barbie paper doll book includes one 9¾" tall paper doll of a blonde Barbie doll and fashions. It originally sold for $1.29. Published by Western Publishing Co. A Golden Book Barbie Paper Doll.

©1985   Uncut $5.00

PERFUME PRETTY PAPER DOLL   #1500

This Perfume Pretty paper doll book includes one 9¾" tall paper doll of a blonde Barbie doll, Ken and Whitney dolls, and fashions. It originally sold in the $1.50 range. Published by Western Publishing Co. A Golden Book Barbie Paper Doll.

©1988   Uncut $5.00

PINK & PRETTY BARBIE PAPER DOLL   #1983-44

This Pink & Pretty Barbie paper doll book includes one 9¾" tall paper doll of a blonde Barbie doll and fashions. It originally sold for $1.29. Published by Western Publishing Co. A Golden Book Barbie Paper Doll.

©1983   Uncut $7.00

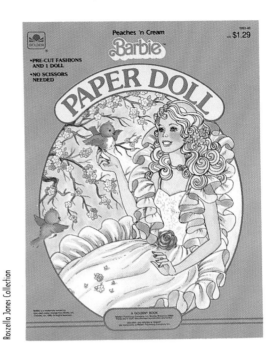

Rozella Jones Collection

Rozella Jones Collection

Rozella Jones Collection

Rozzella Jones Collection

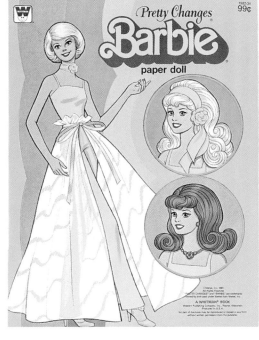

Rozzella Jones Collection

Rozzella Jones Collection

Clockwise from top left:

PINK & PRETTY BARBIE PAPER DOLL PLAYBOOK   **#1836-43**
This Pink & Pretty Barbie paper doll playbook includes one 9¾" tall paper doll of a blonde Barbie doll, fashion show backdrop and fashions. It originally sold for $2.59. Published by Western Publishing Co. A Golden Barbie Paper Doll & Play Book.
©1983   Uncut $10.00

PRETTY CHANGES BARBIE PAPER DOLL   **#1982-34**
This Pretty Changes Barbie paper doll book includes one 9¾" tall paper doll of a blonde Barbie doll and fashions. It originally sold for 99¢. Published by Western Publishing Co. A Whitman Book Barbie Paper Doll.
©1981   Uncut $10.00

BARBIE AND THE ROCKERS PAPER DOLL   **#1528**
This Barbie and the Rockers paper doll book includes one 9¾" tall paper doll of a blonde Barbie doll, Derek, Diva, Dana, and Dee Dee dolls and fashions. It originally sold for $1.29. Published by Western Publishing Co. A Golden Book Barbie Paper Doll.
©1986   Uncut $8.00

Clockwise from top right:

SUNSATIONAL BARBIE PAPER DOLL #1982-44
This Sunsational Barbie paper doll book includes one 9¾" tall paper doll of a blonde Barbie doll, Ken doll, and fashions. It originally sold for $1.29. Published by Western Publishing Co. A Golden Book Barbie Paper Doll.

©1983   Uncut $5.00

SUPERSTAR PAPER DOLL #1537-2
This SuperStar paper doll book includes one 9¾" tall paper doll of a blonde Barbie doll and fashions. It originally sold in the $1.50 range. Published by Western Publishing Co. A Golden Book Barbie Paper Doll.

©1989   Uncut $5.00

TROPICAL PAPER DOLL #1523
This Tropical paper doll book includes one 9¾" tall paper doll of a blonde Barbie doll, Ken, Miko, and Skipper dolls, and fashions. It originally sold for $1.39. Published by Western Publishing Co. A Golden Book Barbie Paper Doll.

©1986   Uncut $5.00

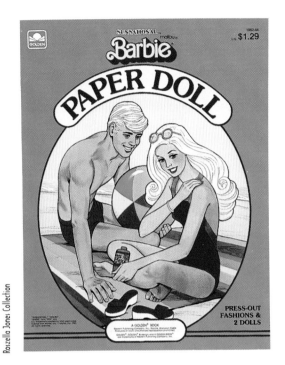

Roszella Jones Collection

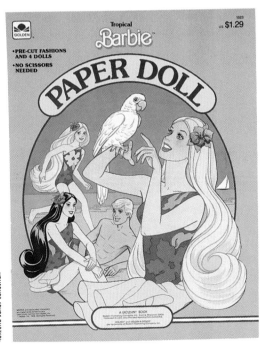

Roszella Jones Collection

Roszella Jones Collection

Roszella Jones Collection

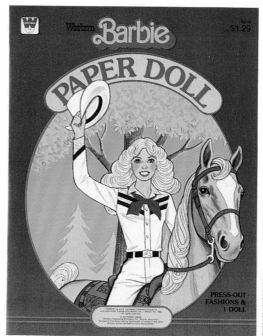

Roszella Jones Collection

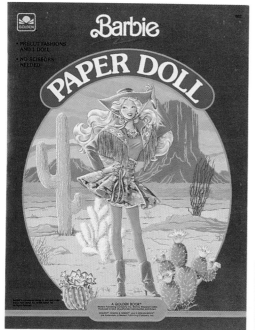

Roszella Jones Collection

Clockwise from top left:

TWIRLY CURLS BARBIE PAPER DOLL #1982-46
This Twirly Curls Barbie paper doll book includes one 9¾" tall paper doll of a blonde Barbie doll and fashions. It originally sold for $1.29. Published by Western Publishing Co. A Golden Book Barbie Paper Doll.
©1983   Uncut $5.00

WESTERN BARBIE PAPER DOLL #1982-43
This Western Barbie paper doll book includes one 9¾" tall paper doll of a blonde Barbie doll and fashions. It originally sold for $1.29. Published by Western Publishing Co. A Whitman Book Barbie Paper Doll. (Back of book also shown.)
©1982   Uncut $7.00

WESTERN BARBIE PAPER DOLL #1502
This Western Barbie paper doll book includes one 9¾" tall paper doll of a blonde Barbie doll and fashions. It originally sold in the $2.00 range. Published by Western Publishing Co. A Golden Book Barbie Paper Doll.
©1990   Uncut $4.00

ANIMALS

Ever since that first pink felt dog that came with Nighty Negligee in 1959, animals have played an engaging part in the Barbie doll collectibles field. When Barbie doll went western, it followed naturally that she would need a horse. Her first horse was Dancer in the 1970s. Some of the horses produced for 1981–1991 were Prancer, Blinking Beauty, Midnight, Champion, Honey, Star Stepper, Sun Runner, Dallas, and Dixie. Barbie doll was portrayed in the role of veterinarian and more domestic pets were produced. The feline Fluff was offered in 1985 and dogs like the Afghan Beauty and her pups, Prince the poodle, and Sachi, a fluffy white dog, came along as some of Barbie doll's favorites. And with the Animal Lovin' promotion, a zebra, giraffe, panda, monkey and lion cub were introduced. All of these animals came with tiny extras in the package, most with grooming tools and instructions. The values for all of these animals are for the NRFB category. (See pricing explanation pages 10–11.) Remember that the average value reported here goes down as the piece is removed from the box, played with, and has parts missing.

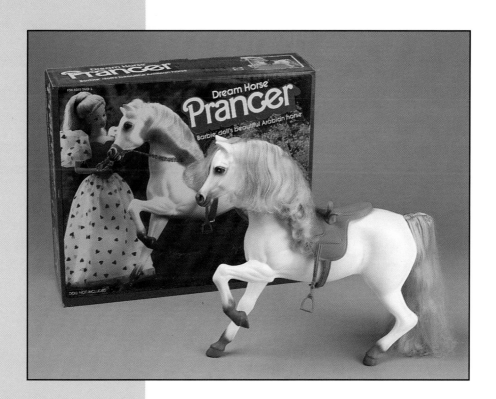

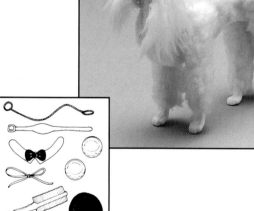

Clockwise from top left:

FLUFF   #5524   1985   NRFB $42.00
Fluff is a white kitten that came with a carrying case, scratching post, bed, mattress, dish, collar, and a how-to care booklet.

BEAUTY AND PUPPIES   #5019   ©1981   NRFB $30.00
An Afghan dog with flowing fur, two puppies, a collar, crown, dog dish, hat, ribbons, and more are included in this package.

PRINCE   #7928   1985   NRFB $30.00
Prince is a glamourous white French poodle. The drawing shows the accessories that come with this pet.

SACHI   #5468   ©1990   NRFB $18.00
This fluffy white all-cloth dog has extras like a scarf, cap, grooming tools, food dish, and more.

Roszella Jones Collection

Roszella Jones Collection

Clockwise from top left:

SNOWBALL   #7272   ©1990   NRFB $18.00
This white furry dog with a formed plastic face comes with grooming tools, food dish, and more.

ANIMAL LOVIN' GINGER THE GIRAFFE   #1395   ©1988   NRFB $37.00
This unique stuffed cloth giraffe is a member of the Animal Lovin' collection.

BARBIE CHAMPION (FOREIGN)   #4045   ©1991   NRFB $38.00
This black horse has a white blaze on its forehead and four white socks on its feet. The box has English as well as foreign languages and was intended for the foreign market. A red plaid saddle blanket, saddle, and red and silver ornaments are a few of the extras with this Barbie horse. Also included in this package is an equestrian outfit for Barbie doll.

Roszella Jones Collection

DALLAS #3312

This is a golden palomino for Barbie and Ken dolls. It was produced over the course of four years beginning just before the time period of this book and extending into it. The box includes a brown saddle, stirrups, bridle with reins, brush, and instruction booklet.

1980   NRFB $32.00

Roszella Jones Collection

DIXIE #7073

Dixie is a palomino and the colt of Dallas. She comes with a pink gingham blanket, brush, birth certificate, halter, lead rein, and instructions.

©1983   NRFB $23.00

Roszella Jones Collection

HONEY #5880

This is Skipper doll's show horse and comes with saddle, bridle, and reins. She is dark brown with a white blaze on her forehead and four white socks.

©1982   NRFB $25.00

Roszella Jones Collection

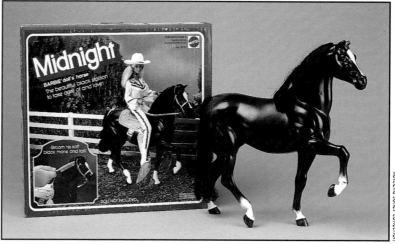

DREAM HORSE MIDNIGHT #5337

This black stallion comes with a silvery bridle, reins, and saddle. Ribbon, stirrups, comb, brush, and instructions are included.

©1983   NRFB $33.00

Roszella Jones Collection

Roszella Jones Collection

Roszella Jones Collection

Author's Collection

Roszella Jones Collection

Clockwise from top left:

DREAM HORSE PRANCER #7263 ©1983 NRFB $33.00
This is an Arabian horse with a silvery mane and tail highlighted in pink that comes with a pink side saddle, comb, ribbons, stickers, and instruction booklet.

ALL AMERICAN STAR STEPPER AND BARBIE #3712 ©1990 NRFB $54.00
This chestnut horse with golden mane and tail is gift packaged with an All American Barbie doll for wholesale clubs. The package includes a special saddle, stirrups, bridle with reins, grooming tools, and more. This is a special value because it includes a doll with the horse.

WESTERN FUN BARBIE AND HER HORSE SUN RUNNER #5408 ©1990 NRFB $65.00
This package contains Barbie doll in the Western Fun outfit, her horse Sun Runner, saddle, stirrups, bridle with reins, blanket, lariat, saddle ornaments, ribbons, and more. This is a special value because it includes a doll with the horse.

ANIMAL LOVIN' ZIZI THE ZEBRA #1393 ©1988 NRFB $37.00
This black and white (of course!) cloth zebra is a member of the Animal Lovin' collection.

ACTIVITY SETS

Activity or play sets were a great way to engage Barbie doll and all the other dolls in fun-time fantasy. While they are not as hot a collectible as the doll itself, they are still very interesting. They also provide a wonderful setting for their corresponding theme dolls. For instance, what could be a more appropriate way to present Barbie and the Rockers dolls than to set them on their Hot Rockin' Stage with their instruments! The detail and imagination utilized in producing these playtime vignettes add a great deal to their popularity. In most instances you will not find these NRFB. But for just a few dollars, you can have coordinated props for displaying your dolls.

Remember that the average value reported here is for NRFB, like-new merchandise. (See pricing explanation pages 10–11.) If the sets are removed or have pieces missing, the value will be much less.

BARBIE 6 O'CLOCK NEWS   #7745
This set has a TV news station backdrop diorama, news desk, camera, control panel, chairs, and more.
©1984   NRFB $45.00

Author's Collection

Print designs with real working stamper!

COOL TOPS SKIPPER FASHION T-SHIRT SHOP #4955
This set includes a folding screen background to hold plain white T-shirts, tiny shelves full of stickers, visors, paint cans, sunglasses, and more. An ink stamper was incorporated into the round counter and different designs could be stamped to stick on the T-shirts.
©1989   NRFB $20.00

Rozella Jones Collection

BARBIE DANCE CLUB DANCETIME SHOP   #4840
This stage set includes jukebox, 6 album covers, radio, 2 guitars, 3 posters, trumpet, cash register, 12 records, 3 microphones, saxophone, magazine, portable keyboard, shelves, racks, counter, labels, and instructions. The jukebox came in yellow and pink. This collection of props can be used with many of the '50s theme dolls in the Doll chapter.
©1989   MNP $45.00

Author's Collection

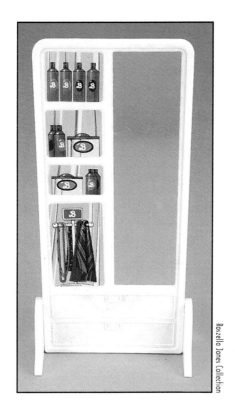

Rozella Jones Collection

Author's Collection

Author's Collection

Clockwise from top left:

BARBIE DREAM STORE MIRROR   #4022
This mirror is from the Barbie Dream Store. The package also included 2 clear plastic display cases, jewelry, glasses, scarfs, furs, hats, purses, chair, and mannequins. The value reported here is for the entire set, not just this mirror. Sold separately these pieces would be no more than a few dollars each.

1984   NRFB   entire set $30.00

CALIFORNIA DREAM BARBIE HOT DOG STAND   #4463
This set is a sandwich stand shaped like a hot dog! This collection came with lots of little extras: bags of Fritos, pretzels and chips, trays, hot dogs, and more! (The top counter for the stand in this photograph has been replaced. Remember that the value reported here is for NRFB.)

1988   NRFB $45.00

BARBIE ICE CREAM SHOPPE   #3653
This set includes a counter, cart that is a real ice cream maker, stools that double as ice cream spoons, and chairs that become ice cream cups. It even has a vinyl pink and teal checked floor for the shoppe!

©1987   NRFB $50.00

ISLAND FUN HUT #4414

This hut has a bright green grass roof and holds a net hammock. A parrot swings on his perch and extras include tropical fruit and shells.

1987　　　　　　NRFB　　　　　$15.00

Roszella Jones Collection

Author's Collection

LIGHTS & LACE BARBIE LIVE CONCERT #8311

This set includes keyboard, tambourine, cymbals, trumpet, drum set, microphone, and more to coordinate with the Lights & Lace Barbie doll series. See the Doll chapter for a photo of this Barbie doll.

©1991　　　　　　NRFB　　　　　$25.00

BARBIE LOVES MCDONALD'S #5559

This set is especially collectible because of the success of the McDonald's promotions. It has the McDonald's logo on the roof of the hamburger stand, a picnic table, fencing, food preparation counter, and lots of tiny extras like food, trays, napkin holder, fries, and more.

©1983　　　　　　NRFB　　　　　$45.00

Roszella Jones Collection

Author's Collection

Rozella Jones Collection

Rozella Jones Collection

Clockwise from top left:

SUPERSTAR BARBIE PIANO CONCERT   #7314
This package includes a white baby grand piano and stool, couch, coffee table, easel, framed photo of Barbie doll, and more accessories. SuperStar Barbie and Ken dolls are shown in the Doll chapter. The piano from this set is featured on the cover.
©1989   NRFB $25.00

BARBIE ELECTRONIC PIANO   #5085
This baby grand piano has 21 working keys and comes with a songbook, 2 playing wands, sheet music for Barbie doll and a piano bench that opens for storage. Battery operated.
©1981   NRFB $33.00

BARBIE PICNIC FUN   #7341
This play set includes a barbecue, cooler, radio, Frisbee, plates, utensils, ball, hot dogs, and hamburgers.
©1989   NRFP $5.00

Roszella Jones Collection

BARBIE POSE ME PRETTY BEAUTY SET　　　　#7161　　　　©1984　　　　NRFB　　　　$20.00
These beauty sets come with a head that would turn with long golden hair to style and curl. They come complete with a tray full of make-up and hair accessories and a brush and comb for hours of fashion fun.

BARBIE AND THE ROCKERS LIVE CONCERT INSTRUMENTS　　#3611　　　　©1986　　　　NRFP　　　　$25.00
This set of musical instruments includes a drum set and drumsticks, trumpet, tambourine, guitar, keyboard, microphone and doll stand. The Barbie and the Rockers promotion produced a large array of collectibles. Be sure to look in the other chapters for more Rockers items.

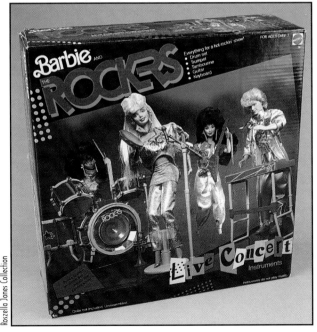

Roszella Jones Collection

Roszella Jones Collection

Roszella Jones Collection

BARBIE AND THE ROCKERS HOT ROCKIN' STAGE   #1144
This set includes 2 keyboards, 2 guitars, stage light, TV monitor, 5 doll stands, "Rockers in Concert" record, a cardboard stage backdrop that on the back was a backstage diorama, 4 dressing room chairs, and 4 comb, brush, and mirror sets.
©1985   NRFP $25.00

BARBIE SODA SHOPPE   #2707
This set includes over 75 pieces. It has a soda fountain that really works, counter, stools that double as child-size soda glasses, and lots more. This collection also is a unique setting for the '50s style of doll or almost any of the casually themed dolls.
©1988   NRFB $38.00

Author's Collection

Roszella Jones Collection

TRACY & TODD WEDDING PLAY PAKS SET   #4936
Thirty-two pieces are included in this set. Two chairs, a table and tablecloth, wedding cake, wedding decorations, punch bowl and glasses plates, knives, forks, spoons, presents, and more make up this wedding day play collection. For corresponding dolls, see Tracy & Todd in the Doll chapter.
©1983   NRFB $20.00

HOUSEHOLD

As with most doll collectors, I have always been fascinated by dollhouses and furniture. My personal preference is for items that are scaled versions of the full-size pieces. For instance, rather than a rectangular box shape for a bed, the 4-poster replica of a full-size version would be my choice. Some examples of furniture and fixtures released for this time period are a working roll-top desk, a canopy bed with a light globe that both revolved and lit up, a working shower, and a bubbling spa. Each of these was either battery operated or had some special functioning feature.

The houses produced over these years were certainly varied. The post-and-beam construction of the three-story town house with a working elevator was used extensively. The backdrop diorama was changed periodically to update this structure. In 1991 the Barbie Magical Mansion, an incredible house with arched windows and two bay windows on the front main floor, was introduced. It featured four white pillars that supported two balconies. It is a wonderful house with lights and even sounds like a real home. Unfortunately, I have no photo of it, but I know you will recognize it when you see it. If you find it like new and still in the box, expect to pay as much as $600.00. Of course if it has been played with, or shows signs of damage or missing parts, the value will be much less.

This is another field where you are not likely to find items NRFB. But, even if they are in played-with condition, they make a perfectly scaled display piece for your dolls. The average values reported here are for NRFB items. (See pricing explanation pages 10–11.) If you find these collectibles out of box or only a piece here and there, the value will be much less.

Rozella Jones Collection

Clockwise from top left:

PINK SPARKLES BARBIE ARMOIRE   **#4763**
This pink armoire has two doors that open and a bottom drawer that pulls out. It is from the Pink Sparkles furniture collection.
©1990   NRFB $20.00

DREAM POOL PATIO BARBECUE   **#1478**
This set includes a barbecue, long-handled utensils, salt and pepper shakers, 2 hot dogs and 2 hamburgers. The barbecue has a transparent cover that lifts off and a grill with sides that open.
©1980   NRFB $10.00

BARBIE GLAMOUR BATH & SHOWER SET   **#2552**   ©1985   NRFP $15.00
Bubble bath, soap, brush, comb, mirror, perfume bottle, back brush, towel rack, hair dryer, shampoo bottle, creme rinse bottle, hair spray container, and towel are the accessories that complete this Barbie Bath & Shower Set.

BARBIE DREAM BED   **#5641**   ©1982   NRFB $25.00
This is a white 4-poster canopy bed and includes pillows, bedspread, sheets, ribbons, label sheet, and cut-outs.

Author's Collection

Clockwise from top:

PINK SPARKLES BARBIE STARLIGHT BED SET   #3739   ©1990   NRFB $30.00

This pink canopy bed has a light globe in the top with star and moon cutouts and a working light. It is from the Pink Sparkles furniture collection. Battery operated.

RIBBONS & ROSES BED SET   #5620   ©1987   NRFB $12.00

The headboard of this bed is reversible with a bow and ribbon on one side and a spray of roses on the other.

BARBIE BUBBLE BATH   #5280   ©1981   NRFP $25.00

This package includes a sunken garden tub, working shower, bubble bath, brush, soap, towel, vanity chair, and more.

Roszella Jones Collection

Roszella Jones Collection

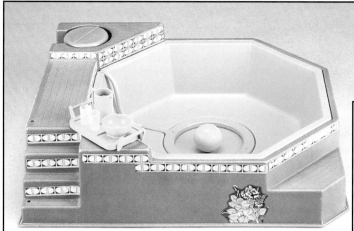

Author's Collection

BARBIE BUBBLING SPA #7145
This spa can be filled with water and, by use of the push pump at the corner, it will actually bubble! It comes with a beach ball, towel, tray, bowl, glasses, and pitcher. It was introduced in 1984.
©1984   Mint $25.00

Author's Collection

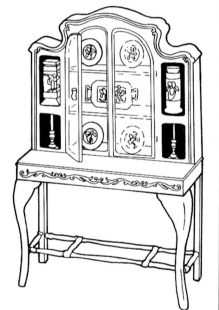

BARBIE DREAM FURNITURE BUFFET #9479
This buffet with clear plastic doors is from the Dream Furniture collection. It came with extras of plates, tray, candlesticks, vases, and more.
1985   NRFB $25.00

Roxzella Jones Collection

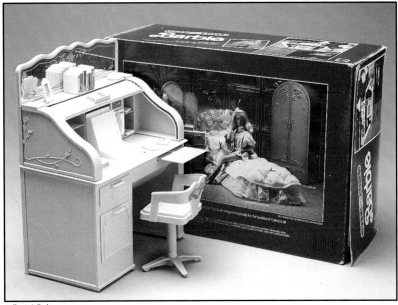

Author's Collection

Clockwise from top left:

SWEET ROSES BARBIE COOKING CENTER   #4777
This stove, oven, and microwave cabinet set comes with silvery pots and pink pans and utensils. It is in the very popular Sweet Roses line of furniture.
1987   NRFB $20.00

SWEET ROSES BARBIE ROLL TOP DESK   #3639
This pink roll top desk complete with accessories and chair is in the very popular Sweet Roses line of furniture. On the back of the box, the bed for this line of furniture is featured.
©1990   NRFB $25.00

BARBIE WASH & WATCH DISHWASHER   #2232
This kitchen countertop is complete with dishwasher and sink that really work. The dishes change color when they are washed. It is made to coordinate with the Sweet Roses and Pink Sparkles furniture collections.
©1991   NRFB $20.00

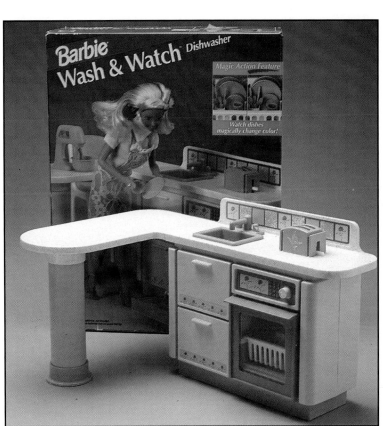

Author's Collection

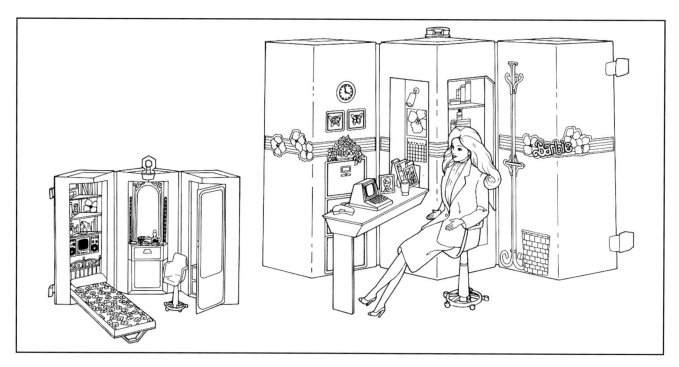

BARBIE HOME & OFFICE SET   #7895   1985   NRFB $20.00

This is a 3-section hard plastic case that opens to reveal a bedroom with vanity, bed, shelves, and closet on one side. On the reverse side is an office vignette with desk, phone, computer, and more. While I have no photo, this line drawing should identify this set enough for you to recognize it when seen.

BARBIE DREAM HOUSE   #2587   1985   NRFB $200.00

This house style was popular and produced in this orange, yellow, and white color and in pink and white in later years. It is in 3 sections and can be placed in several arrangements. It comes with extra flooring for patio areas and plastic plants for window boxes.

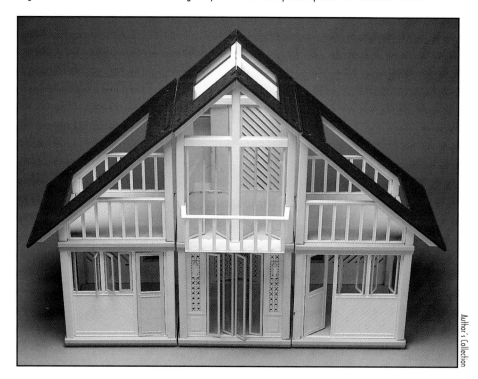

BARBIE GLAMOUR HOME #9477

New in 1985, this two-story house for Barbie doll has a spiral staircase at one end and an attached canopy that held a swing at the other. The rooftop is a patio and furniture is included. While I have no photo, this line drawing should identify the structure enough for you to recognize it.

©1984   NRFB $95.00

BARBIE TOWNHOUSE #9042

This three-story, 6 room townhouse comes complete with furniture and working elevator. This structure design was a consistent best-seller with the diorama backdrop being changed from time to time. This backdrop has photographs of room settings from actual homes and this diorama was first released in 1984.

©1989   NRFB $80.00

Roszella Jones Collection

BARBIE DREAM KITCHEN   #9119

This kitchen complex includes a stove with oven, sink, refrigerator (under the cabinet), trash compactor, and it comes with an assortment of accessories. This set contains more than 80 fabulous pieces.

©1984   Mint $25.00

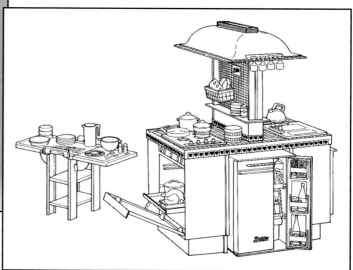

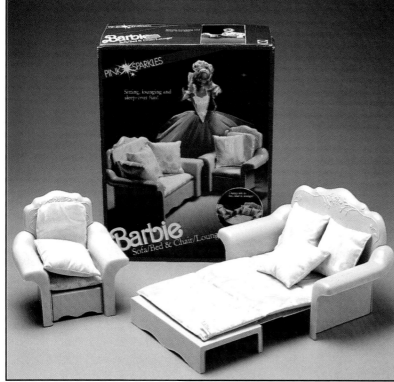

PINK SPARKLES BARBIE SOFA/BED AND
CHAIR/LOUNGER   #4771

This pink sofa and chair combination comes with cushions made from white material printed with pastel pink, teal, and blue accented with glitter. It is from the Pink Sparkles furniture collection.

©1990   NRFB $20.00

Author's Collection

Author's Collection

BARBIE FASHION LIVING ROOM SET   #7404
This wicker-looking set includes a sofa/bed, chair/lounger, coffee table, end table, cushions, cups, saucers, teapot, tray, and phone.
1985   NRFB $35.00

Roszella Jones Collection

BEACH BLAST POOL & PATIO SET   #3593
This set includes 58 pieces and the hard plastic pool actually holds water.
1987   NRFB $25.00

CALIFORNIA DREAM BARBIE POOL PARTY   #7762
Made by Arco, this set contains a vinyl pool, 2 lounge chairs, table, potted plant, beach ball, utensils, plates, glasses, soft drinks, apples, banana, oranges, pineapple, and grapes.
©1987   NRFB $15.00

Roszella Jones Collection

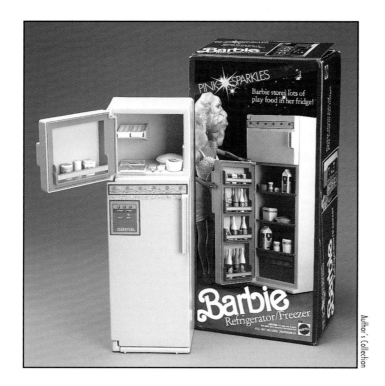

Author's Collection

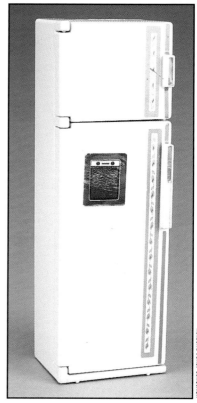

Roszella Jones Collection

Author's Collection

Clockwise from top left:

PINK SPARKLES BARBIE REFRIGERATOR/FREEZER   #4776
This pink refrigerator has doors that open and shelves to hold food. It is from the Pink Sparkles furniture collection.
©1990   NRFB $22.00

SWEET ROSES BARBIE REFRIGERATOR   #4776
This refrigerator has trim featuring rosebuds. It is in the very popular Sweet Roses line of furniture.
1987   NRFB $22.00

PINK SPARKLES BARBIE FUN PHONE CENTER   #1707
This pink table has leaves that lift and lock in place at each end. It also has a phone that rings and speaks and lamp that lights. It is from the Pink Sparkles furniture collection and is battery operated.
©1990   NRFB $20.00

BARBIE DREAM FURNITURE TABLE & CHAIRS #9480

This table and chairs are from the Dream Furniture collection. It has reversible cushions and comes with only two chairs. Extras include dinnerware and a center lazy susan.

1985   NRFB $25.00

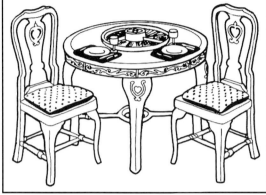

Author's Collection

SWEET ROSES BARBIE DINING TABLE & CHAIRS   #7107

This pink dining table and set of four chairs come with reversible cushions, placemats, candelabrum, plates, goblets, and other tableware. The table top is also reversible.

©1987   NRFB $25.00

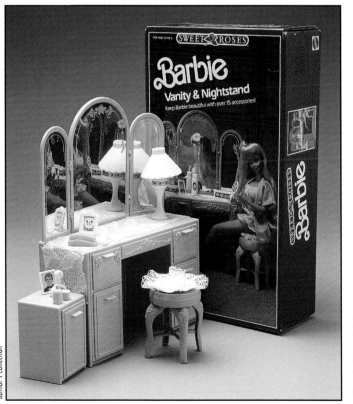

Author's Collection

Author's Collection

SWEET ROSES BARBIE VANITY & NIGHTSTAND   #4764

This pink vanity with three-way mirror and nightstand comes with a stool, cushion, lace dresser scarf, and 13 beauty accessories. It is in the very popular Sweet Roses line of furniture.

©1987   NRFB $25.00

SWEET ROSES BARBIE 3-PIECE WALL UNIT #4772

This pink wall unit has clear doors on the top and side shelves and is in the very popular Sweet Roses line of furniture. It comes with 15 play pieces to fill its shelves.
©1987   NRFB $22.00

Author's Collection

PINK SPARKLES BARBIE WASHER & DRYER #1706

Suds appear in the window of this pink washer. It has a matching dryer and is from the Pink Sparkles furniture collection. Battery operated.
©1990   NRFB $25.00

Rozella Jones Collection

BARBIE WORKOUT CENTER LOCKER #7975

This locker is a part of a set named Workout Center. It comes with this piece and exercise cycle, slant board, weight machine, and extras. Remember, the item by itself is worth only a few dollars. It must be complete for the value given here.
1985   NRFB $20.00 for the set

PAPER GOODS

Ephemera is by definition an item produced for use only once or for a short time. So most of this collection is paper ephemera. Not as popular as the rest of the collection arenas, but certainly entertaining, stickers, napkins, cards, postcards, calendars, party goods, tissues, and posters are just a few of these unique items. Paper goods are pleasing and reflect the art and fashion of the comparable time period and should be fairly inexpensive for this interval of years. Values are for unused merchandise — still in its original packaging or in like-new condition. (See pricing explanation pages 10–11.) If the items are opened, used, or showing wear, the average value reported here will be much less.

Roszella Jones Collection

Roszella Jones Collection

Author's Collection

Clockwise from top left:

1989–AMERICA'S FAVORITE DOLL BARBIE CALENDAR
This calendar was produced by Gibson Greetings Inc. It was designed and photographed by Tom Rovito and featured Barbie doll modeling many of the 900 Series of fashions.
©1988   Mint $15.00

1991–BARBIE WEEKLY CALENDAR
This small pocket calendar was produced by Gibson Greetings Inc.
©1990   Mint $2.00

1991–NOSTALGIC BARBIE CALENDAR
This calendar was produced by Gibson Greetings Inc. and featured Barbie dolls wearing the 900 Series of fashions.
©1990   Mint $10.00

Roxzella Jones Collection

Roxzella Jones Collection

Roxzella Jones Collection

Clockwise from top left:

1991–SNEAK PREVIEW CALENDAR
This was a 15-month calendar that was packaged with a Barbie doll fashion and rebate coupons.
©1990　　　　　　NRFP　　　　　$12.00

BARBIE SHOPPING SPREE CARD GAME　　　　#4888
This is a Golden Giant card game produced by Western Publishing Co., Inc.
©1991　　　　　　NRFP　　　　　$3.00

8 BARBIE THANK YOU CARDS AND ENVELOPES
Produced by Reed Paper, these 3½" x 5" thank you cards feature Barbie doll in her special Happy Birthday gown.
©1989　　　　　　NRFP　　　　　$3.00

Rozella Jones Collection

Rozella Jones Collection

8 BARBIE THANK YOU CARDS AND ENVELOPES #1292
Produced by Unique Industries, Inc., Philadelphia, PA, these 3¼" x 5½" thank you cards feature Barbie with a heart and roses.
©1989   NRFP $3.00

8 BARBIE INVITATIONS #1284   NRFP $3.00
Produced by Unique Industries, Inc., Philadelphia, PA, these 3¼" x 5½" party invitations feature Barbie doll with a vanity mirror.

BARBIE PLAYING CARDS DECK
This is a set of 52 playing cards and 2 jokers. Lots of fashions are pictured throughout these cards. This deck could be ordered through the Barbie Pink Stamp Club for $1.75 on the club membership form.
1990   Mint $3.00

Rozella Jones Collection

BARBIE CENTERPIECE #1289
Produced by Unique Industries Inc., this cardboard and tissue honeycomb decoration could be used on a cake or as a centerpiece for a birthday party.
©1988   NRFP $5.00

BARBIE SHHH–BARBIE AND FRIEND AT PLAY DOOR HANGER
Made by the Antioch Publishing Company, this door hanger features art work of Barbie doll in a strapless party dress.
©1988   Mint $2.00

Shhh–
Barbie and friend at play

Rozella Jones Collection

BARBIE WELCOME DOOR HANGER ©1991 Mint $2.00
Made by the Antioch Publishing Co., this door hanger features a close-up of Barbie doll surrounded by roses.

BARBIE FAN CLUB PACKAGE ©1981 Mint $8.00
This package comes in a bright pink envelope and contains games, puzzles, note pad, coupons, newsletter, and poster to color.

BARBIE GIFT BAG
1991 Mint $1.50
This gift bag is white with a pink bottom and bright accents with the Barbie logo on the front.

BARBIE GIFT WRAP & CARD ©1988 NRFP $10.00
This gift wrap features Barbie doll and fashions to cut out and use on a blue background. A birthday card was also produced to coordinate with this paper.

BARBIE PLATES
Produced by Unique Industries Inc., these 9" diameter plastic coated plates feature Barbie doll in an off-one-shoulder dress with hearts and flowers in the background.
©1985   NRFP $5.00

BARBIE NAME TAGS
Produced by Paper Art, these 3¼" x 2⅛" stick-on name tags feature Birthday Barbie doll in her pink birthday gown.
©1990   NRFP   pkg. $2.00

BARBIE LUNCHEON NAPKINS
Produced by Unique Industries Inc., these 6½" square napkins feature Barbie doll with a perfume bottle and hearts and flowers in the background.
©1988   NRFB $3.00

NOSTALGIC BARBIE NAPKINS
Produced by C.A. Reed, Inc., these 5" square napkins feature the Nostalgic Barbie logo.
©1990   NRFP   pkg. $3.00

Roszella Jones Collection

Roszella Jones Collection

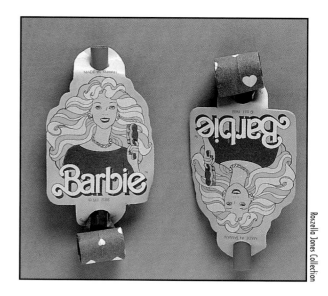

Roszella Jones Collection

BARBIE PARTY FAVOR
This blower party favor features a cardboard card with Barbie doll holding a brush.
©1988   Mint $2.00

BARBIE PARTY FAVORS
This party favor package contains a set of six personal note pads with the artwork of Barbie doll on the front.
©1988   NRFP   pkg. $3.00

Roszella Jones Collection

BARBIE POSTCARDS
Great photographs of Barbie doll through the years were produced in these sets of postcards manufactured by the American Postcard Company of New York. They were sold at specialty, gift, card, and doll shops.
1989   Mint   80¢ each

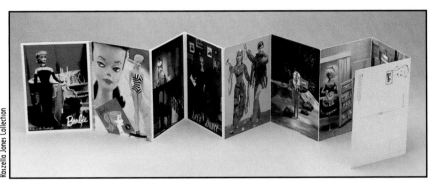
Roszella Jones Collection

BARBIE THROUGHOUT HISTORY POSTCARDS
This is a set of 9 cards featuring Barbie dolls from 1959 through 1988. They came folded together and could be detached and used.
1989   Mint   set $5.00

30TH ANNIVERSARY POSTER
This 30th anniversary poster shows 900 Series Barbie fashions and is 22½" x 28½".
©1989   Mint $10.00

Roszella Jones Collection

Roszella Jones Collection

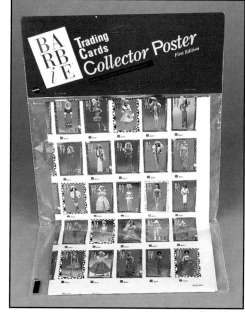

Author's Collection

Clockwise from top left:

DREAM ROOM POSTER
This poster is designed with a space for someone to write in her own name to display on bedroom door or wall.
©1989   Mint $4.00

TRADING CARDS COLLECTOR POSTER   #5529
This poster shows all 300 Barbie fashion trading cards and is 23½" x 46".
©1990   Mint $10.00

CREATIVE TEACHING PRESS BARBIE STICKERS   Rose-#5133
This is a package of rose-scented, scratch 'n sniff paper Barbie stickers.
©1983   NRFP $2.00

CREATIVE TEACHING PRESS BARBIE STICKERS   Gardenia-#5135
This is a package of gardenia-scented scratch 'n sniff paper Barbie stickers.
©1983   NRFP $2.00

Roszella Jones Collection

Roszella Jones Collection

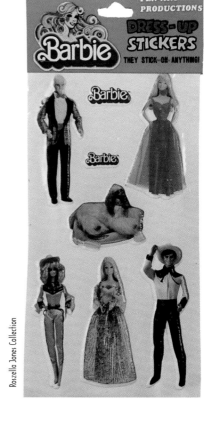

Clockwise from top right:

PEN–ART BARBIE STICKERS   #175
This is a package of dimensional vinyl Barbie stickers.
©1981   NRFP $3.00

REED PAPER ART BARBIE STICKERS   #0426-025
This is a package of paper Barbie Shoppin' Spree stickers.
©1990   NRFP $1.50

REED PAPER ART BARBIE STICKERS   #0437-025
This is a package of paper Birthday Barbie stickers.
©1990   NRFP $1.50

Roszella Jones Collection

Roszella Jones Collection

Roszella Jones Collection

BARBIE TISSUES
This package contains six smaller blue, purple, and pink packs of tissues, each featuring characters from the Barbie line.
©1988   NRFP   pkg. $2.00

BARBIE TRADING CARDS 10 CARD PACK   #5528
This retail store display of 10 card packs contains 24 packs making the entire display and contents worth $48.00.
©1990   NRFP   per pack $2.00

BARBIE TRADING CARDS GIFT PACK   #5528
Forty trading cards featuring mini stories on the back of each card. Also issued was a complete boxed set of 300 trading cards valued at $60.00 for the set in perfect shape.
©1990   NRFP $8.00

PUBLICATIONS

This category contains reading, coloring, and activity books as well as magazines and comic books. Most were produced with the intent of entertaining and stimulating children and so contain whimsical artwork. Childhood memories are often tied to a favorite book kept close at hand and constantly reviewed and reread. Little Golden Books is a separate collectible field by itself. To a Barbie doll collector, a Little Golden Book about Barbie doll is doubly collectible. Of course, the most valuable Little Golden books are from the 1940s, long before Barbie doll was produced.

Naturally, *Barbie Bazaar* is most in demand in this field. It is a wonderful compilation of the historical progression of Barbie doll aimed toward collectors. All of these collectibles contain a treasure trove of nostalgic information. Books and periodicals other than *Barbie Bazaar* will not be in a high value bracket in this time period because so many were produced. (See pricing explanation pages 10–11.) As with all fields, condition of the item will strongly affect the average value reported here.

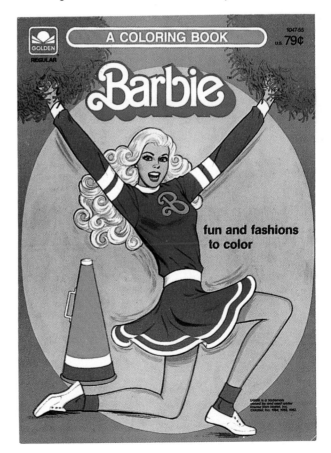

Roszella Jones Collection

Clockwise from top left:

ADVENTURES WITH BARBIE #4

This is produced by Price Stern Sloan. It is a story book named "Soda Shop Surprise" with a few line drawings included.
©1991   Mint $4.00

BARBIE THIRTY YEARS OF AMERICA'S DOLL

Notes and black and white photos of Barbie doll year by year are featured in this book by Cynthia Robins, published by Contemporary Books, 128 pages, 8" x 5".
©1989   Mint $9.00

NOSTALGIC BARBIE – A POSTCARD BOOK

Produced by the American Postcard Company and published by Running Press Book Publishers, contains 30 postcards, 6¾" x 4¾".
©1990   Mint $16.00

FOREVER BARBIE – A POSTCARD BOOK

Produced by the American Postcard Company and published by Running Press Book Publishers, contains 30 postcards, 6¾" x 4¾".
©1991   Mint $16.00

BARBIE BAZAAR MAGAZINE PREMIERE EDITION

This premiere edition of *Barbie Bazaar* Magazine features a #2 ponytail Barbie doll in Gay Parisienne. The original plans were to publish 10 times per year.
1988   Mint $75.00

Roszella Jones Collection

Roszella Jones Collection

Roszella Jones Collection

BARBIE BAZAAR MAGAZINE SEPTEMBER 1988
This edition of *Barbie Bazaar* Magazine features a cover with a rare black Francie doll being crowned Miss Teenage Beauty by Alan doll with a brunette Francie in Sweet 'N Swingin', and a blonde Francie in First Formal looking on.
1988   Mint $45.00

Roszella Jones Collection

Roszella Jones Collection

BARBIE BAZAAR MAGAZINE OCTOBER 1988
This edition of *Barbie Bazaar* Magazine features a cover with the artwork of Dick Tahsin. It is a drawing of a blonde Bubble Cut doll wearing Enchanted Evening. The back featured a unique ad with replicas of Maba Company fashions from the early 1960s in the surroundings of Rockefeller Plaza in New York. This ad was on the next six issues.
1988   Mint $45.00

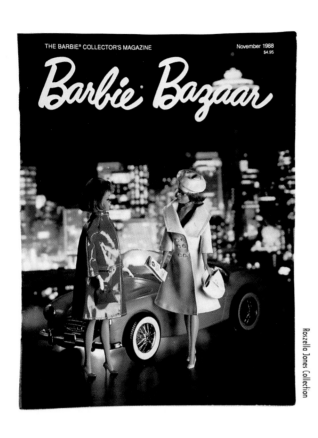

Roszella Jones Collection

Clockwise from top left:

BARBIE BAZAAR MAGAZINE NOVEMBER 1988

This edition of *Barbie Bazaar* Magazine features a cover with a brownette American Girl in Fashion Shiner and a red-haired bendable leg Midge doll in London Tour.

| 1988 | Mint | $45.00 |

BARBIE BAZAAR MAGAZINE DECEMBER 1988

This edition of *Barbie Bazaar* Magazine features a cover with a brunette American Girl Barbie doll in Benefit Performance gown. A change in the original plans to publish 10 times per year was announced in this issue. From the next issue forward, they were produced bi-monthly.

| 1988 | Mint | $55.00 |

BARBIE BAZAAR MAGAZINE JANUARY/FEBRUARY 1989

This edition of *Barbie Bazaar* Magazine features a cover with a photo of a #1 blonde ponytail doll meant to mimic the cover of that first pink fashion booklet.

| 1989 | Mint | $20.00 |

Roszella Jones Collection

Roszella Jones Collection

Clockwise from top left:

BARBIE BAZAAR MAGAZINE MARCH/APRIL 1989
This edition of *Barbie Bazaar* Magazine features a cover with artwork by Mel Odom of Fashion Queen Barbie doll.
1989   Mint $20.00

BARBIE BAZAAR MAGAZINE MAY/JUNE 1989
This edition of *Barbie Bazaar* Magazine features a cover with a #5 brunette ponytail doll dressed in Wedding Day set tossing her bouquet to Skipper doll dressed as Junior Bridesmaid.
1989   Mint $20.00

BARBIE BAZAAR MAGAZINE JULY/AUGUST 1989
This edition of *Barbie Bazaar* Magazine features a cover with Beach Blast Barbie doll leaning against her purple metallic Corvette at the beach.
1989   Mint $20.00

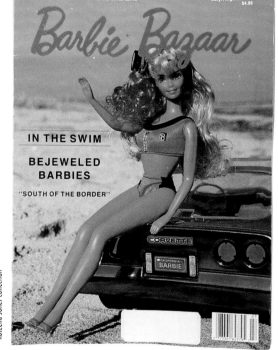

Rozella Jones Collection

Rozella Jones Collection

Rozella Jones Collection

Clockwise from top left:

BARBIE BAZAAR MAGAZINE SEPTEMBER/OCTOBER 1989
This edition of *Barbie Bazaar* Magazine features a cover with a #1 Barbie doll in Easter Parade.

1989	Mint	$20.00

BARBIE BAZAAR MAGAZINE NOVEMBER/DECEMBER 1989
This edition of *Barbie Bazaar* Magazine features a cover with a Swirl ponytail doll in Red Flair accented for Christmas.

1989	Mint	$23.00

BARBIE BAZAAR MAGAZINE JANUARY/FEBRUARY 1990
This edition of *Barbie Bazaar* Magazine features a cover with a Bubble Cut doll wearing the widely recognizable Solo in the Spotlight.

1990	Mint	$18.00

Clockwise from top right:

BARBIE BAZAAR MAGAZINE MARCH/APRIL 1990
This edition of *Barbie Bazaar* Magazine features a cover with a twist 'n turn doll with real eyelashes and a bouquet of daisies in the foreground.
1990   Mint $18.00

BARBIE BAZAAR MAGAZINE MAY/JUNE 1990
This edition of *Barbie Bazaar* Magazine features a cover with a Bubble Cut Barbie doll in Senior Prom and Ken doll in tuxedo.
1990   Mint $18.00

BARBIE BAZAAR MAGAZINE JULY/AUGUST 1990
This edition of *Barbie Bazaar* Magazine features a cover with a Living Barbie doll dressed in Bloom Bursts holding an umbrella. Her unique earrings for this photo-shoot were made by Barry Sturgill.
1990   Mint $18.00

Rozella Jones Collection

Clockwise from top left:

BARBIE BAZAAR MAGAZINE SEPTEMBER/OCTOBER 1990
This edition of *Barbie Bazaar* Magazine features a cover with four side-part Barbie dolls dressed in four different formals.
| 1990 | Mint | $18.00 |

BARBIE BAZAAR MAGAZINE NOVEMBER/DECEMBER 1990
This edition of *Barbie Bazaar* Magazine features a cover with a Color Magic Barbie doll in a Christmas setting at the North Pole.
| 1990 | Mint | $18.00 |

BARBIE BAZAAR MAGAZINE JANUARY/FEBRUARY 1991
This edition of *Barbie Bazaar* Magazine features a cover with Air Force Barbie doll.
| 1991 | Mint | $16.00 |

Rozella Jones Collection

Rozella Jones Collection

Roxzella Jones Collection

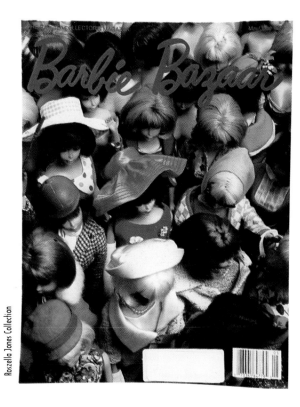

Roxzella Jones Collection

Clockwise from top left:

BARBIE BAZAAR MAGAZINE MARCH/APRIL 1991
This edition of *Barbie Bazaar* Magazine features a cover with Mel
Odom's portrait of a Sleep-eyed Miss Barbie doll.
1991   Mint $16.00

BARBIE BAZAAR MAGAZINE MAY/JUNE 1991
This edition of *Barbie Bazaar* Magazine features a cover with a crowd
of American Girls in 1600 Series fashions.
1991   Mint $16.00

BARBIE BAZAAR MAGAZINE JULY/AUGUST 1991
This edition of *Barbie Bazaar* Magazine features a cover with a Bubble
Cut Barbie doll and Ken doll in wedding attire.
1991   Mint $16.00

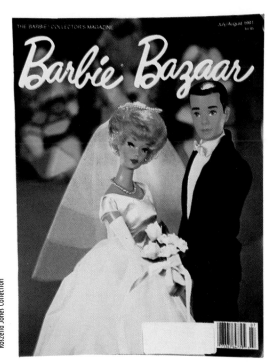

Roxzella Jones Collection

Rozella Jones Collection

Rozella Jones Collection

Clockwise from top left:

BARBIE BAZAAR MAGAZINE SEPTEMBER/OCTOBER 1991
This edition of *Barbie Bazaar* Magazine features a cover with Barbie and Ken dolls in college attire with a backdrop sold through Sears in the mid '60s.
1991   Mint $16.00

BARBIE BAZAAR MAGAZINE NOVEMBER/DECEMBER 1991
This edition of *Barbie Bazaar* Magazine features a cover with a unique doll with gold hair wearing the Lamé Pak Sheath from 1963. It was from the collection of Glenn Offield.
1991   Mint $16.00

PINK & PRETTY BIG COLORING BOOK   **#1147-21**
This Western Publishing Co., Inc. Golden coloring book features Pink & Pretty Barbie doll on the cover. It originally sold for 99¢.
©1983   Mint $3.00

THE *BARBIE 30TH ANNIVERSARY MAGAZINE*
This special little girls' magazine was produced in celebration of the 30th anniversary.
One of the great features is the two-page spread of the Barbie family tree.
©1989   Mint $15.00

Roszella Jones Collection

Roszella Jones Collection

BALLERINA BARBIE COLOR/ACTIVITY BOOK   #5522-1
This Western Publishing Co., Inc. deluxe color and activity book features
Barbie doll as a ballerina on the cover. It has a Barbie paper doll on the
back cover.
©1990   Mint $2.00

Roszella Jones Collection

Rozella Jones Collection

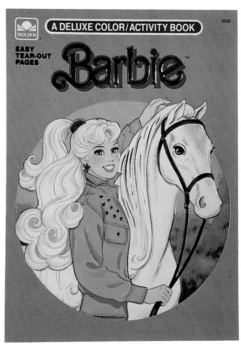

Rozella Jones Collection

Rozella Jones Collection

Clockwise from top left:

BARBIE AND THE ROCKERS COLOR/ACTIVITY BOOK   #5510
This Western Publishing Co., Inc. deluxe Barbie color and activity book features Barbie and the Rockers dolls on the cover.
©1987   Mint $2.50

BARBIE COLOR/ACTIVITY BOOK   #5522
This Western Publishing Co., Inc. deluxe color and activity book features Barbie doll with a white horse on the cover. It has a Barbie paper doll on the back cover.
©1988   Mint $2.50

BARBIE AND THE ROCKERS BIG COLORING BOOK   #1147–32
This Western Publishing Co., Inc. Golden coloring book features Barbie and the Rockers dolls on the cover. It originally sold for 99¢.
©1986   Mint $2.50

Clockwise from top right:

BARBIE AND THE ROCKERS GIANT COLORING BOOK #3105–86
This Western Publishing Co., Inc. Golden coloring book features Barbie and the Rockers dolls performing on the cover. It originally sold for $1.49.
©1986   Mint $2.50

BARBIE BIG COLORING BOOK   #1146–21
This Western Publishing Co., Inc. Golden coloring book features Barbie doll and her horse on the cover with a blue background. It originally sold for 99¢.
©1985   Mint $3.00

BARBIE LOVING YOU BIG COLORING BOOK   #1146–24
This Western Publishing Co., Inc. Golden coloring book features Loving You Barbie doll holding a rose on the cover. It originally sold for 99¢.
©1985   Mint $3.00

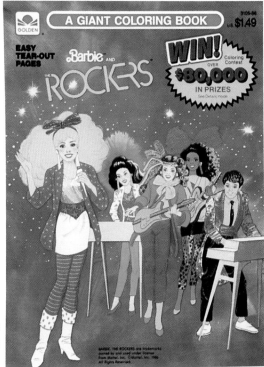

Clockwise from top left:

BARBIE BIG COLORING BOOK #1147-85
This Western Publishing Co., Inc. Golden coloring book features Barbie doll and her dog Prince on the cover with a blue background. It originally sold for 99¢.
©1985 Mint $3.00

BARBIE COLORING BOOK #1829-34
This Western Publishing Co., Inc. Golden coloring book features Barbie doll at a bus stop on the cover. It originally sold for $1.29.
©1983 Mint $3.00

BARBIE COLORING BOOK #1146-6
This Western Publishing Co., Inc. Golden coloring book features Barbie doll and her horse Dallas on the cover. It originally sold for 99¢.
©1984 Mint $3.00

Roszella Jones Collection

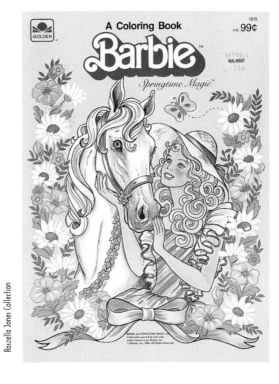

Roszella Jones Collection

Clockwise from top left:

BARBIE COLORING BOOK #1147-70
This Western Publishing Co., Inc. Golden coloring book features Barbie doll and her dream horse Prancer on the cover. It originally sold for 99¢.
©1984   Mint $3.00

BARBIE COLORING BOOK #1315
This Western Publishing Co., Inc. Golden coloring book features Barbie doll and her horse on the cover with a peach background. It originally sold for 99¢.
©1985   Mint $3.00

BARBIE GIANT COLORING BOOK #3175-1
This Western Publishing Co., Inc. Golden coloring book features Barbie doll looking through a photo album on the cover. It originally sold for $1.49.
©1988   Mint $2.50

Roszella Jones Collection

Clockwise from top left:

BARBIE GIANT COLORING BOOK #1146–33
This Western Publishing Co., Inc. Golden coloring book features Barbie doll on a pink to white to purple background cover.
©1990 Mint $2.00

BARBIE & KEN COLORING BOOK #1047–58
This Western Publishing Co., Inc. coloring book features Barbie and Ken dolls on a roller-coaster on the cover. It originally sold for 79¢.
©1986 Mint $2.50

BARBIE & KEN GIANT COLORING BOOK #1829–42
This Western Publishing Co., Inc. Golden coloring book features Barbie doll receiving a rose from Ken doll on the cover. It originally sold for $1.29.
©1985 Mint $3.00

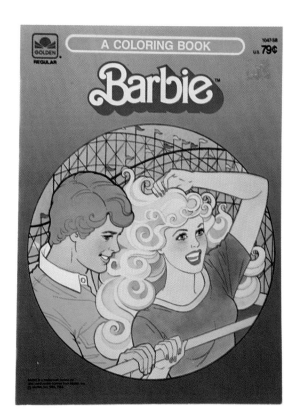

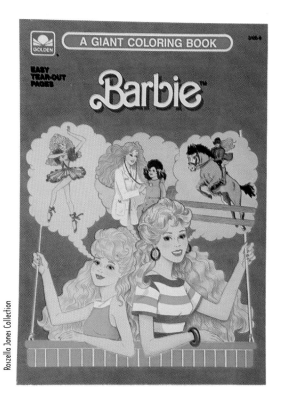

Clockwise from top right:

BARBIE & SKIPPER GIANT COLORING BOOK   #3105-9
This Western Publishing Co., Inc. Golden coloring book features Barbie and Skipper dolls imagining activities on the cover.
©1989   Mint $2.00

CALIFORNIA DREAM BIG COLORING BOOK   #1146-8
This Western Publishing Co., Inc. Golden coloring book features California Dream Barbie doll on the cover.
©1988   Mint $2.50

CHEERLEADER BARBIE COLORING BOOK   #1047-55
This Western Publishing Co., Inc. Golden coloring book features Cheerleader Barbie doll on the cover. It originally sold for 79¢.
©1984   Mint $3.00

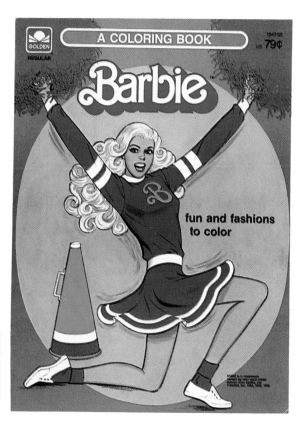

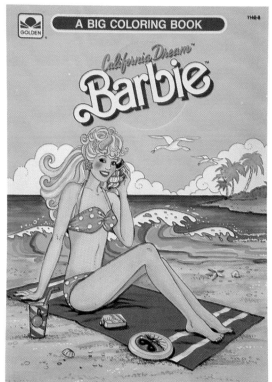

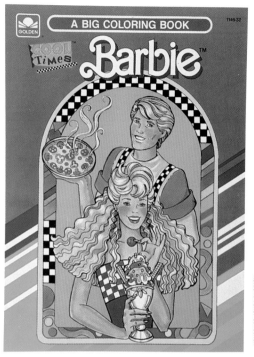

Clockwise from top left:

COOL TIMES BIG COLORING BOOK   #1146-32
This Western Publishing Co., Inc. Golden coloring book features Cool Times Barbie and Ken dolls on the cover.
©1989   Mint $2.00

CRYSTAL BARBIE BIG COLORING BOOK   #1146-23
This Western Publishing Co., Inc. Golden coloring book features Crystal Barbie and Ken dolls on the cover. It originally sold for 99¢.
©1985   Mint $3.00

CRYSTAL BARBIE COLORING BOOK   #1147-23
This Western Publishing Co., Inc. Golden coloring book features Crystal Barbie and Ken dolls on the cover. It originally sold for 99¢.
©1984   Mint $3.00

Roszella Jones Collection

Clockwise from top left:

DAY-TO-NIGHT BIG COLORING BOOK   #1146–85

This Western Publishing Co., Inc. Golden coloring book features Day-To-Night Barbie doll on the cover. It originally sold for 99¢.

©1985   Mint $3.00

DREAM DATE BARBIE COLORING BOOK   #1147–22

This Western Publishing Co., Inc. coloring book features Dream Date Barbie and Ken dolls on the cover. It originally sold for 99¢.

©1983   Mint $3.00

DREAM GLOW BIG COLORING BOOK   #1147–28

This Western Publishing Co., Inc. Golden coloring book features Dream Glow Barbie and Ken dolls in a rose arbor on the cover. It originally sold for 99¢.

©1986   Mint $2.50

Roszella Jones Collection

Roszella Jones Collection

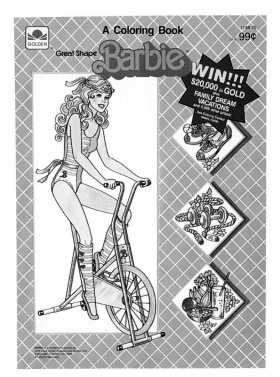

Rozella Jones Collection

GREAT SHAPE BARBIE COLORING BOOK   #1146-70
This Western Publishing Co., Inc. Golden coloring book features Great Shape Barbie doll on the cover. It originally sold for 99¢.
©1984   Mint $3.00

LOVING YOU BIG COLORING BOOK   #1147-30
This Western Publishing Co., Inc. Golden coloring book features Loving You Barbie doll on the cover. It originally sold for 99¢.
©1984   Mint $3.00

PEACHES 'N CREAM BIG COLORING BOOK   #1146-11
This Western Publishing Co., Inc. Golden coloring book features Peaches 'n Cream Barbie doll on the cover. It originally sold for 99¢.
©1985   Mint $3.00

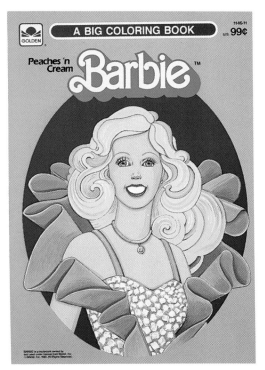

Rozella Jones Collection

Rozella Jones Collection

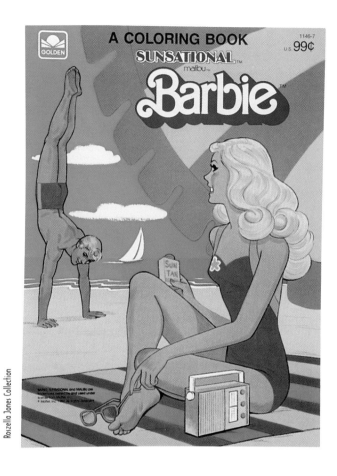

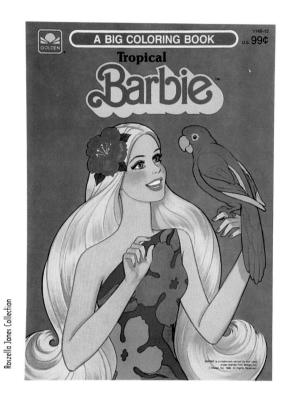

Roszella Jones Collection

Roszella Jones Collection

Clockwise from top left:

SUNSATIONAL MALIBU BARBIE COLORING BOOK   **#1146-7**
This Western Publishing Co., Inc. Golden coloring book features Sunsational Malibu Barbie and Ken dolls on the cover. It originally sold for 99¢.
©1984   Mint $3.00

TROPICAL BIG COLORING BOOK   **#1146-12**
This Western Publishing Co., Inc. Golden coloring book features Tropical Barbie doll on the cover. It originally sold for 99¢.
©1986   Mint $2.50

WEDDING PARTY BARBIE COLORING BOOK   **#1047-56**
This Western Publishing Co., Inc. coloring book features Barbie and Skipper Wedding Party dolls on the cover. It originally sold for 79¢.
©1985   Mint $3.00

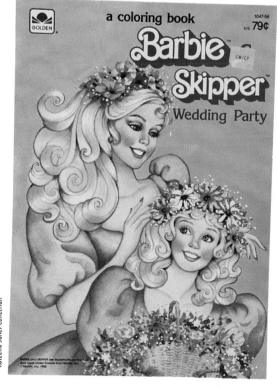

Roszella Jones Collection

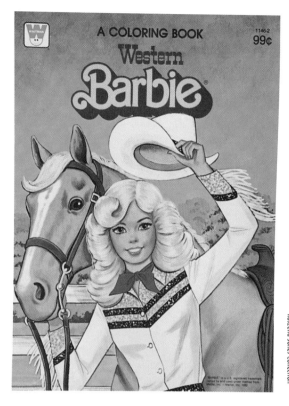

Roxzella Jones Collection

Roxzella Jones Collection

Roxzella Jones Collection

Clockwise from top left:

WESTERN BARBIE COLORING BOOK   **#1146-2**
This Western Publishing Co., Inc., Whitman coloring book features Western Barbie doll on the cover. It originally sold for 99¢.
©1982   Mint $3.50

WESTERN BIG COLORING BOOK   **#1218-4**
This Western Publishing Co., Inc. Golden coloring book features Western Barbie doll in Santa Fe holding a southwestern pot on the cover.
©1990   Mint $2.00

WESTERN BIG COLORING BOOK   **#1218-5**
This Western Publishing Co., Inc. Golden coloring book features Western Barbie doll on a trip to Santa Fe on the cover.
©1990   Mint $2.00

Clockwise from top right:

BARBIE COMICS #1 JANUARY

Marvel Comics began in this year to produce Barbie and Barbie Fashion comic books featuring stories, craft activities, and letters to Barbie. As a rule, the first in a series of comic books is always worth more. These comics will continue to go up in price but at this date, they have not increased a great amount. This first issue Barbie doll comic came packaged with a Barbie doll doorknob hanger.

©1991   Mint $5.00

BARBIE COMICS #2 FEBRUARY

Marvel Barbie comic book. This cover features Barbie and Ken dolls traveling in a horse drawn sleigh through a snowy scene.

©1991   Mint $2.00

BARBIE COMICS #3 MARCH

Marvel Barbie comic book. This cover features Barbie and Ken dolls dancing the night away!

©1991   Mint $2.00

Rozella Jones Collection

Clockwise from top left:

BARBIE COMICS #4 APRIL
Marvel Barbie comic book. This cover features Barbie doll in high fly-ing fashion fun.
©1991   Mint $2.00

BARBIE COMICS #5 MAY
Marvel Barbie comic book. This cover features a variation of SuperStar Barbie doll on a blue starry background.
©1991   Mint $2.00

BARBIE COMICS #6 JUNE
Marvel Barbie comic book. This cover features Barbie and Ken dolls dancing to the top of the pops on a record with a red musical back-ground.
©1991   Mint $2.00

Rozella Jones Collection

Rozella Jones Collection

Clockwise from top right:

BARBIE COMICS #7 JULY
Marvel Barbie comic book. This cover features Barbie doll on a tropical beach in a pink and white bathing suit.
©1991   Mint $2.00

BARBIE COMICS #8 AUGUST
Marvel Barbie comic book. This cover features Barbie doll in jump jeans.
©1991   Mint $2.00

BARBIE COMICS #9 SEPTEMBER
Marvel Barbie comic book. This cover features Barbie doll and friends on a ferris wheel.
©1991   Mint $2.00

Rozella Jones Collection

Rozella Jones Collection

Rozella Jones Collection

Clockwise from top left:

BARBIE COMICS #10 OCTOBER
Marvel Barbie comic book. This cover features Barbie and Skipper dolls working out at the gym.
©1991   Mint $2.00

BARBIE COMICS #11 NOVEMBER
Marvel Barbie comic book. This cover features Barbie doll in jeans and black and red shirt seated on a fence in a fall scene.
©1991   Mint $2.00

BARBIE COMICS #12 DECEMBER
Marvel Barbie comic book. This cover features Barbie doll walking 101 (well, almost) Dalmatians!
©1991   Mint $2.00

Roszella Jones Collection

Clockwise from top left:

BARBIE FASHION COMIC & BARBIE COMIC GIFT SET

Marvel Comics began in this year to produce Barbie and Barbie Fashion comic books featuring stories, craft activities, and letters to Barbie doll. This is a gift set including both first issues and a free gift.

©1991   Mint $10.00

BARBIE FASHION COMICS #1 JANUARY

Marvel Comics began in this year to produce Barbie and Barbie Fashion comic books featuring stories, craft activities, and letters to Barbie doll. As a rule, the first in a series of comic books is always worth more. These comics will continue to go up in price but at this date, they have not increased a great amount. This first issue Fashion comic came packaged with a Barbie Comic Pink Card.

©1991   Mint $5.00

BARBIE FASHION COMICS #2 FEBRUARY

Marvel Barbie Fashion comic book. This cover features Barbie doll in a black and white outfit on a black and white checkerboard background.

©1991   Mint $2.00

BARBIE FASHION COMICS #3 MARCH

Marvel Barbie Fashion comic book. This cover features Barbie doll sitting on a stool reading a stack of valentines in front of a white heart on a red background.

©1991   Mint $2.00

Roszella Jones Collection

Rozella Jones Collection

Rozella Jones Collection

Rozella Jones Collection

Clockwise from top left:

BARBIE FASHION COMICS #4 APRIL
Marvel Barbie Fashion comic book. This cover features Barbie doll and the big New York City adventure.
©1991   Mint $2.00

BARBIE FASHION COMICS #5 MAY
Marvel Barbie Fashion comic book. This cover features Barbie doll and friends in Celebrate Spring With Sensational Syles.
©1991   Mint $2.00

BARBIE FASHION COMICS #6 JUNE
Marvel Barbie Fashion comic book. This cover features Barbie doll in a blue and white starry outfit with a red and white stripe background.
©1991   Mint $2.00

Clockwise from top right:

BARBIE FASHION COMICS #7 JULY
Marvel Barbie Fashion comic book. This cover features Barbie doll celebrating Skipper doll's birthday.
©1991   Mint $2.00

BARBIE FASHION COMICS #8 AUGUST
Marvel Barbie Fashion comic book. This cover features Barbie doll in a flower print dress with the flower print continued in the background.
©1991   Mint $2.00

BARBIE FASHION COMICS #9 SEPTEMBER
Marvel Barbie Fashion comic book. This cover features Barbie doll and friends at the beach under blue skies and umbrellas.
©1991   Mint $2.00

Rozella Jones Collection

Clockwise from top left:

BARBIE FASHION COMICS #10 OCTOBER
Marvel Barbie Fashion comic book. This cover features Barbie doll in a black and white fashion with a black beret.
©1991   Mint $2.00

BARBIE FASHION COMICS #11 NOVEMBER
Marvel Barbie Fashion comic book. This cover features Barbie doll curled up in a cushioned chair reading "My Memory Book."
©1991   Mint $2.00

BARBIE FASHION COMICS #12 DECEMBER
Marvel Barbie Fashion comic book. This cover features Barbie doll in an island fashion surrounded by exotic birds, plants, and flowers.
©1991   Mint $2.00

Rozella Jones Collection

Rozella Jones Collection

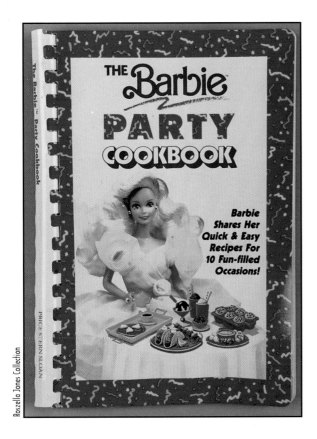

The Barbie™ Party Cookbook

PRICE STERN SLOAN

Roxella Jones Collection

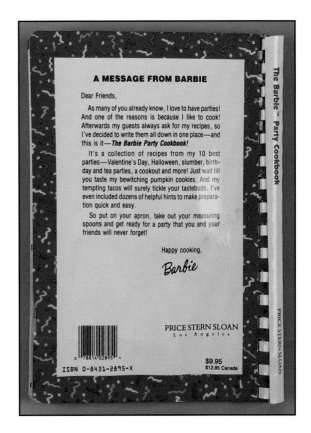

A MESSAGE FROM BARBIE

Dear Friends,

As many of you already know, I love to have parties! And one of the reasons is because I like to cook! Afterwards my guests always ask for my recipes, so I've decided to write them all down in one place—and this is it—*The Barbie Party Cookbook!*

It's a collection of recipes from my 10 best parties—Valentine's Day, Halloween, slumber, birthday and tea parties, a cookout and more! Just wait till you taste my bewitching pumpkin cookies. And my tempting tacos will surely tickle your tastebuds. I've even included dozens of helpful hints to make preparation quick and easy.

So put on your apron, take out your measuring spoons and get ready for a party that you and your friends will never forget!

Happy cooking,

Barbie

PRICE STERN SLOAN
Los Angeles

ISBN 0-8431-2895-X

$9.95
$12.95 Canada

The Barbie™ Party Cookbook

PRICE STERN SLOAN

Clockwise from top left:

BARBIE PARTY COOKBOOK
This cookbook was produced by Price Stern Sloan and originally sold for $9.95. It is rare to find one that does not have kitchen spills and signs of use!

      Mint $12.00

BARBIE DIARY
This Western Publishing Co., Inc. Golden Barbie diary was written partially by Barbie with blanks for a child to use.

©1985   Mint $5.00

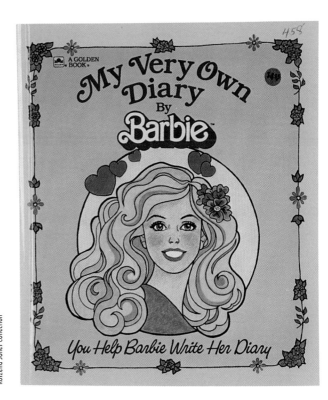

A GOLDEN BOOK

My Very Own Diary By Barbie™

You Help Barbie Write Her Diary

Roxella Jones Collection

Roszella Jones' Collection

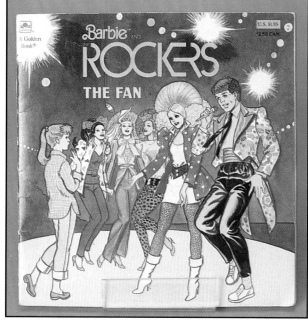

Author's Collection

BARBIE IN DREAM VACATION 8003 ©1984   Mint $4.00
A Listen 'n Look Book by Hasbro Industries, Inc.

BARBIE AND THE ROCKERS THE FAN ©1987   Mint $3.00
A Little Golden Book, Western Publishing Co., Inc.

A PICNIC SURPRISE   ©1990   Mint $2.00
A Little Golden Book, Western Publishing Co., Inc.

THE MISSING WEDDING DRESS   ©1986   Mint $5.00
A Little Golden Book, Western Publishing Co., Inc.

Author's Collection

Author's Collection

Rozzella Jones Collection

Clockwise from top left:

MATTEL ANNUAL REPORT
This is a summary report for the stockholders of Mattel for 1988.
1988   Mint   No Price Available

MATTEL ANNUAL REPORT
This is a summary report for the stockholders of Mattel for 1990.
1990   Mint   No Price Available

MATTEL ANNUAL REPORT
This is a summary report for the stockholders of Mattel for 1991.
1991   Mint   No Price Available

Rozzella Jones Collection

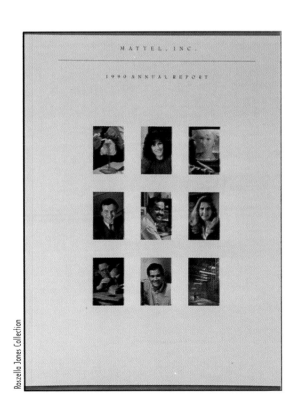

Rozzella Jones Collection

1983 SPRING
WHITE SPACE SALE

Rozzella Jones Collection

Mattel Toys 1984

Rozzella Jones Collection

Mattel Toys 1985

Rozzella Jones Collection

Clockwise from top left:

SPRING AHEAD WITH MATTEL 1983 SPRING SALE CATALOG
This is a saddlestitched summary of a limited amount of toys.
©1982   Mint   No Price Available

MATTEL TOYS 1984 CATALOG
This wonderful wholesale catalog is filled with Mattel's releases for 1984. The first section features Barbie doll, and lots more toys are included. A great resource for information, but for only a limited selection of Barbie items.
©1984   Mint   No Price Available

MATTEL TOYS 1985 CATALOG
This wonderful wholesale catalog is filled with Mattel's releases for 1985. The first section features Barbie doll, and lots more toys are included. A great resource for information, but for only a limited selection of Barbie items.
©1985   Mint   No Price Available

Roszella Jones Collection

Clockwise from top left:

MATTEL 1985 REPRO ART BOOK
This is a saddlestitched collection of line art for use in advertising for Barbie doll and many more Mattel dolls and toys.
©1985   Mint   No Price Available

GOLDEN BARBIE PAINT WITH WATER BOOK   #1785-92
Western Publishing Co., Inc. produced this paint-with-water book featuring Barbie doll holding a first place quilt on the cover.
©1983   Mint $3.00

GOLDEN BARBIE PAINT WITH WATER BOOK   #1785-2
Western Publishing Co., Inc. produced this paint-with-water book featuring Island Fun Barbie doll on the cover.
©1988   Mint $3.00

Roszella Jones Collection

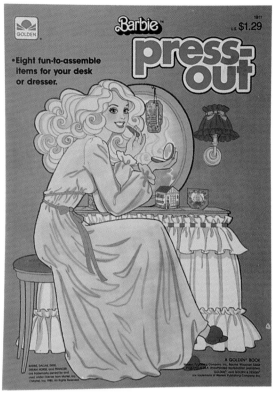

Clockwise from top left:

BARBIE PRESS-OUT BOOK   #1911
This Western Publishing Co., Inc. Golden press-out book features Barbie
doll in a pink nightgown at her dressing table.
©1985   Mint $2.50

GOLDEN BARBIE & THE ROCKERS STICKER FUN BOOK   #2338
Western Publishing Co., Inc. produced these sticker fun books containing 4 pages of stickers and 16 picture pages. The back has a draw and
color activity.
©1987   Mint $2.00

GOLDEN BARBIE STICKER FUN BOOK   #2135-1
Western Publishing Co., Inc. produced these sticker fun books containing 4 pages of stickers and 16 picture pages. The back has a cut-out
item on the back.
©1983   Mint $3.00

Clockwise from top right:

PANINI BARBIE STICKER ALBUM

Panini U.S.A. Co., Inc. distributed these sticker albums. They came with only a portion of the stickers needed. The rest had to be ordered from the Panini Co.

©1983   Mint $2.00

PANINI BARBIE STICKER ALBUM

Panini U.S.A. Co., Inc. distributed these sticker albums. They came with only a portion of the stickers needed. The rest had to be ordered from the Panini Co.

©1989   Mint $2.00

PANINI BARBIE STICKERS

This is a package of 30 stickers that could be ordered from the Panini Co. to complete the Panini Barbie Sticker album.

©1983   Mint $2.00

Roszella Jones Collection

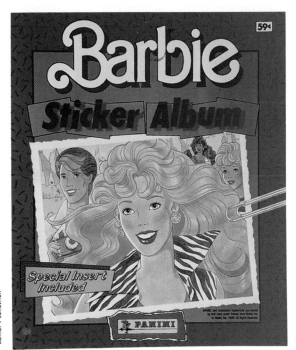

Author's Collection

Roszella Jones Collection

Roszella Jones Collection

Clockwise from top left:

TOPPS BARBIE STICKER ALBUM
Topps Chewing Gum, Inc. distributed these sticker albums. They came with only a portion of the stickers needed. The rest had to be ordered from the Topps Co. The back cover features an ad for a child-size Barbie Dreamcycle from Murray®.
©1983   Mint $2.00

TOPPS BARBIE STICKERS
This is a package of 36 Topps stickers that could be ordered to complete the Topps Barbie Sticker album.
©1983   Mint $8.00

BARBIE & THE ROCKERS TRACE & COLOR #2115-2
This Western Publishing Co., Inc. Golden trace & color book features Barbie and the Rockers dolls on the cover. It originally sold for 99¢.
©1986   Mint $2.50

Roszella Jones Collection

MISCELLANEOUS

There are so many articles produced for the Barbie doll collecting market that the categories are indefinable, bringing the word "miscellaneous" to mind to cover the limitless categories. From Action Accents for Barbie doll's home to a video featuring Barbie and the Rockers dolls, this is a fun area in which to find Barbie doll items. China tea sets are available and have been popular for many years. Mattel has made these sets child size for little girls' play.

Barbie doll has also had many forms of transportation. Automobiles have been made using popular sports cars of the time period. The one that stirred the most nostalgic memories has to be the '57 Chevy, scale sized for Barbie doll. Avon produced little girls' jewelry and grooming items. Travel cases were sold to carry the dolls and wardrobes from house to house.

On Barbie's thirtieth anniversary commemorative medallions, and even a three-dimensional store display sign were produced. Puzzles and patterns, miniatures and McDonald's figures, and even towel sets all add up to make this field of collecting uniquely interesting. As with all the categories, condition dictates value. (See pricing explanation pages 10–11.) To pay for or sell items at the average value reported here, these items must be NRFB, NRFP, or mint and complete.

Rozzella Jones Collection

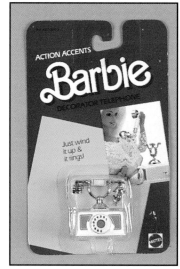

Rozzella Jones Collection

Clockwise from top left:

ACTION ACCENTS BARBIE CLOCK　　　　　#7936　　　　©1989　　　NRFP　　　$3.00
This is one of nine items in the 1989 Action Accents line. When wound, the hands on the clock move and the pendulum swings.

ACTION ACCENTS BARBIE SEWING MACHINE　　#1983　　　　©1986　　　NRFP　　　$3.00
This is one of six items in the 1986 Action Accents line. The machine's needle goes up and down when wound.

ACTION ACCENTS BARBIE DECORATOR TELEPHONE　#1980　　　　©1986　　　NRFP　　　$3.00
This is one of six items in the 1986 Action Accents line. The telephone makes a ringing sound when wound.

'57 CHEVY　　　　　　　　　　　　　#3561　　　　©1989　　　NRFB　　　$75.00
This classic '57 Chevrolet convertible was produced in blue and pink. It has working wheels and the trunk opens for storage. It is "the coolest car in town!" and is a very desirable addition to any collection.

Author's Collection

Roszella Jones Collection

DREAM'VETTE #3299
This pink Corvette comes with stickers to detail the sporty car for Barbie doll.
©1982 NRFB $48.00

FERRARI
This is a white 328 GTS Ferrari. The wheels roll and a special Barbie license plate is included.
1986 NRFB $45.00

Roszella Jones Collection

Roszella Jones Collection

FERRARI
This is a glossy red reproduction of the 328 GTS Ferrari. The wheels roll and a special Barbie license plate is included.
1987 NRFB $45.00

JEEP
This pink Jeep has wheels that roll and was produced to coordinate with the Animal Lovin' promotion. Remember that the price given is for this item complete and in its original packaging.
©1987 NRFB $14.00

Roszella Jones Collection

Author's Collection

PORSCHE #1084

This stylish Porsche is a convertible with working headlights! It is magenta with a white interior and a play cellular phone is detachable from its cradle. The headlight switch is in the console and operates on 2 "C" batteries.

©1991 NRFB $50.00

SILVER'VETTE #4934

This silver Corvette comes with stickers to detail the sporty car for Barbie doll. The hatchback, rear window and glove compartment open and close.

1984 NRFB $38.00

Roszella Jones Collection

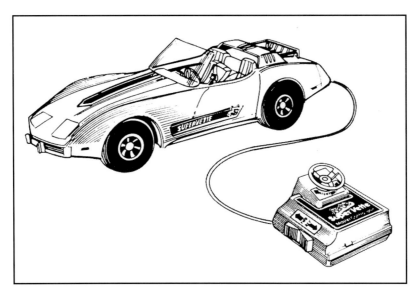

SUPER'VETTE #1291

This special Corvette comes with stickers for detailing and was fashioned after the Sting Ray Corvette. It is special because it can be operated by remote control. A separate control box could steer and make the car move forward or backward. It operates on 2 "D" batteries. While I have no photo, this line drawing should identify the car enough for you to recognize it.

©1981 NRFB $48.00

BARBIE BRACELET/AVON

This is one in a set produced by Avon. It is a 6½" bracelet of pink, dark pink, and clear beads with a heartshape bead at the center.

©1989   NRFB $4.00

BARBIE EARRINGS/AVON

This is one in a set produced by Avon. It includes a set of heartshape pierced earrings.

©1989   NRFB $3.00

BARBIE NECKLACE/AVON

This is one in a set produced by Avon. It is a 15" necklace of pink, dark pink, and clear beads with three heartshape beads included.

©1989   NRFB $5.00

Roszella Jones Collection

Roszella Jones Collection

BARBIE COLOGNE/AVON

This is one in a set produced by Avon. It is a 1.9 oz. bottle of a light, pretty fragrance.

©1989   NRFB $2.00

BARBIE GROOMING SET/AVON

This is one in a set produced by Avon. It includes a Barbie brush and comb.

©1989   NRFB $10.00

Roszella Jones Collection

Roszella Jones Collection

Roszella Jones Collection

Roszella Jones Collection

Roszella Jones Collection

Clockwise from top left:

BARBIE SILK-SCREENED BALLOONS
Manufactured for Reed™ by Paper Art, there are six 11" balloons in the package. They are purple, pink, and blue with white lettering and designs silk-screened on them.
©1990   NRFP $2.00

BARBIE BEAN-BAG
Produced by Lewco Corp., this is a child size hot pink bean-bag chair with artwork of Barbie doll holding a bird on her hand and a dog in the bottom left corner.
©1990   Mint $15.00

BARBIE HAPPY BIRTHDAY CANDLE   #1290
Produced by Unique Industries, Inc., Philadelphia, PA, this candle features Barbie doll with a heart and roses.
©1988   NRFP $4.00

BARBIE FOR GIRLS CAP
Barbie for Girls offered this purple and pink cap produced by Marvel, a Division of Universal Industries, Inc.
©1991   Mint $3.00

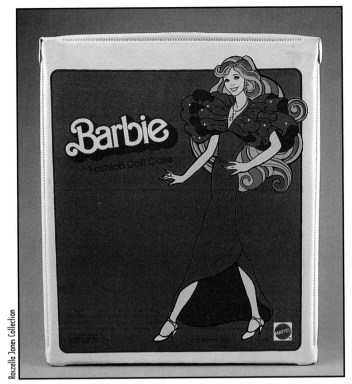

Clockwise from top left:

BARBIE FASHION DOLL CASES   **#1002**
These vinyl-coated cases with Velcro closure feature artwork of a Barbie doll in a hot pink gown. Each is a single doll case. These photos show two versions.
©1982   Mint $8.00

BARBIE FASHION DOLL CASE   **#2810**
This vinyl-coated case with Velcro closure is solid pink with an overall pattern of tiny silver "B"s. There is a Barbie logo in silver at the bottom right corner. It is a single doll case.
©1985   Mint $5.00

Rozella Jones Collection

Rozella Jones Collection

Rozella Jones Collection

Rozella Jones Collection

Clockwise from top left:

BARBIE FASHION DOLL CASE   #2070-9991   ©1988   Mint $8.00
This vinyl-coated case is double sided and closes with a plastic toggle closure on the front. It features artwork of two display windows with Barbie doll in a long pink evening dress in one, and a short daytime dress in the other. This photo shows the back of the case which is for a single doll.

GOLDEN DREAM BARBIE FASHION DOLL CASE   #1002   ©1980   Mint $9.00
This vinyl-coated case with Velcro closure features artwork of Golden Dream Barbie doll on a pink background. It is a single doll case.

MY FIRST PRINCESS BARBIE PLAYCASE   #12020   ©1989   NRFB $5.00
This vinyl-coated case features My First Princess Barbie doll photo on the cover and has a plastic toggle closure.

PINK & PRETTY DOLL CASE   #8320   ©1982   Mint $8.00
This vinyl-coated case has three cardboard drawers. It has a yellow flower border with artwork of Barbie doll, the Barbie logo on the cover, and a metal latch closure.

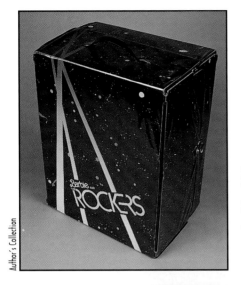

Author's Collection

Roszella Jones Collection

Roszella Jones Collection

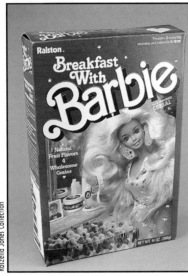

Roszella Jones Collection

Clockwise from top left:

BARBIE AND THE ROCKERS FASHION DOLL CASE　　　　　　　©1985　　　Mint　　　$20.00
This vinyl-coated case with metal clasp closure is black with pink, orange, and lime splatters, spotlight beams, and the Rocker lettering in lime. This case will carry 2 dolls and their fashions.

BARBIE AND THE ROCKERS CASSETTE PLAYER　　　　　　　1986　　　Mint　　　$15.00
This turquoise plastic, battery-operated cassette player was made to coordinate with the Rockers promotion.

BREAKFAST WITH BARBIE BRAND SWEETENED CEREAL　　　　©1989　　　Mint　　　$6.00
Ralston® produced boxes with three different front covers. This one features Dance Club Barbie doll.

BREAKFAST WITH BARBIE BRAND SWEETENED CEREAL　　　　©1989　　　Mint　　　$6.00
Ralston® produced this cereal and if you can find a box, you have a rarity. The back of the box could be cut and folded to make a vanity table for Barbie doll. This box features SuperStar Barbie doll.

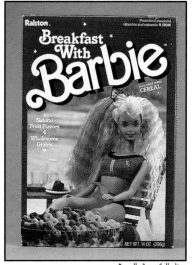

Roszella Jones Collection

Roszella Jones Collection

Roszella Jones Collection

Roszella Jones Collection

Clockwise from top left:

BREAKFAST WITH BARBIE BRAND SWEETENED CEREAL
Ralston® produced boxes with three different front covers. This one features Beach Blast Barbie doll.
©1989   Mint $6.00

BARBIE QUARTZ TALKING ALARM CLOCK
This pink plastic, battery-operated clock has Barbie and Ken dolls posing in front of the clock face.
©1983   Mint $15.00

BARBIE QUARTZ TALKING ALARM DISCO CLOCK
This peach plastic, battery-operated clock has artwork of Barbie and Ken dolls dancing on the plastic lens of the clock face.
©1983   Mint $15.00

BARBIE AND THE ROCKERS DRESS UP SET
This set includes a Stand Up–Dress Up Rocker Barbie and Colorforms fashions that stick and can be removed like magic.
©1986   NRFB $4.00

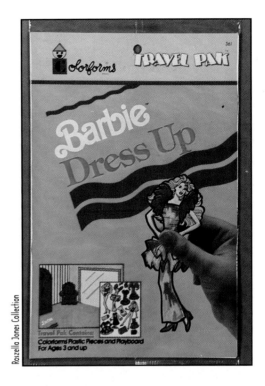

Clockwise from top left:

BARBIE COLORFORMS® DRESS UP SET ©1983 NRFB $4.00
This set includes a Stand Up–Dress Up Barbie doll and Colorforms fashions that stick and can be removed like magic.

BARBIE COLORFORMS® TRAVEL PAK #361 ©1989 NRFB $2.00
This travel pak has a playboard, Barbie doll, and Colorforms fashions that stick and can be removed like magic.

BARBIE 16 CRAYONS ©1983 NRFP $3.00
Produced by the Western Publishing Co., Inc., this is a set of 16 crayons in a clear plastic box.

BARBIE 16 CRAYONS WITH SHARPENER ©1989 NRFB $2.00
Produced by the Western Publishing Co., Inc., this set of 16 crayons is in a spill-proof white plastic box with a sharpener in the bottom.

Rozella Jones Collection

Rozella Jones Collection

Rozella Jones Collection

Clockwise from top left:

BARBIE COLOR & WIPE-OFF BOARD   #2771-1   ©1985   NRFP $4.00
Produced by the Western Publishing Co., this set of 6 crayons comes with a board that can be wiped clean and reused.

BARBIE LONGER LASTING CRAYONS   #50550   ©1990   NRFP $8.00
Produced by the Craft House Corporation, this set of 5 crayons has the Barbie logo stamped in the crayon on one side, and Barbie doll's face on the other. The colors are blue, flesh, red, yellow, and pink.

BARBIE FLAG BANNER   ©1989   Mint $3.00
Produced by Reed Paper, this is a banner of plastic, scallop-edge triangular flags for a birthday celebration.

BARBIE PRETTY SURPRISE HAND CREAM   ©1991   Mint $2.00
This bottle has a large, plastic, shiny magenta bow that slips onto the screw-on top. It is from the line of Barbie Pretty Surprise cosmetics that were sold separately. As the years pass, items like this and many other seemingly disposable items will be harder and harder to find.

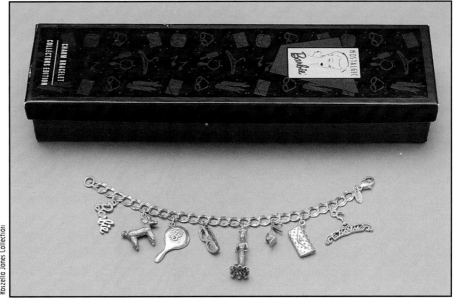

Roszella Jones Collection

BARBIE CHARM BRACELET

Designed by Randi Johnson and manufactured by Peter Brams Designs under license from Mattel, this was a part of the 1989 nostalgic program. The price listed is for the sterling silver version. The gold plate version is valued at $99.00.

1989   NRFP $150.00

Roszella Jones Collection

BARBIE PRINCESS PUNCH

A three-pack of 100% all natural fruit juice punch. Each juice box contains 8.45 oz. Distributed by Natural Kids Foods, Inc. Are you surprised that I didn't find a value for this item! If you would like to report one, please contact me!

©1990   NRFP   No Price Available

BARBIE KITE

Produced as Spectra Star, this kite features artwork of a roller skating Barbie doll at the beach printed on lightweight plastic film supported by a plastic framework.

©1988   Mint $5.00

Roszella Jones Collection

BARBIE & THE ROCKERS LUNCH BOX

This magenta plastic lunchbox by Thermos features artwork of Barbie and the Rockers dolls. A thermos bottle is included.

©1987   Mint $10.00

Roszella Jones Collection

Roszella Jones Collection

Clockwise from top left:

1991 MCDONALD BARBIE AND HOT WHEEL DISPLAY
The first Barbie/McDonald promotion was wildly successful. Collectors had to scramble to make sure they had all eight pieces. Even people from my office were eating fun meals because I bugged them to help me out!
1991   Mint $98.00

1991 MCDONALD BARBIE FIGURINES
This is the 1991 McDonald plastic figurine replicating Lights & Lace Barbie doll.
1991   NRFP $3.50 each

1991 MCDONALD BARBIE FIGURINES
This is the 1991 McDonald plastic figurine replicating Hawaiian Barbie doll.
1991   NRFP $3.50 each

1991 MCDONALD BARBIE FIGURINES
This is the 1991 McDonald plastic figurine replicating Ice Capades Barbie doll.
1991   NRFP $3.50 each

Roszella Jones Collection

Author's Collection

Roszella Jones Collection

Author's Collection

Author's Collection

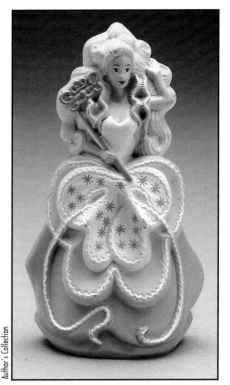

Author's Collection

Clockwise from top left:

1991 MCDONALD BARBIE FIGURINES
This is the 1991 McDonald plastic figurine replicating My First Princess Barbie doll.
1991   NRFP $3.50 each

1991 MCDONALD BARBIE FIGURINES
This is the 1991 McDonald plastic figurine replicating Wedding Day Barbie doll.
1991   NRFP $3.50 each

1991 MCDONALD BARBIE FIGURINES
This is the 1991 McDonald plastic figurine replicating Costume Ball Barbie doll.
1991   NRFP $3.50 each

1991 MCDONALD BARBIE FIGURINES
This is the 1991 McDonald plastic figurine replicating All American Barbie doll.
1991   NRFP $3.50 each

1991 MCDONALD BARBIE FIGURINES
This is the 1991 McDonald plastic figurine replicating Happy Birthday Barbie doll.
1991   NRFP $3.50 each

Author's Collection

Author's Collection

Clockwise from top left:

30TH ANNIVERSARY COMMEMORATIVE MEDALLION

This medallion was produced in sterling silver to celebrate the 30th year of the Barbie doll. Some collectors refer to these commemoratives as coins but unless issued by a mint for use as currency, they should be designated as medallions.

1989	Mint	No Price Available

30TH ANNIVERSARY COMMEMORATIVE MEDALLION

These medallions were produced in silver and gold finish to celebrate the 30th year of the Barbie doll. Some collectors refer to these commemoratives as coins but unless issued by a mint for use as currency, they should be designated as medallions.

1989	Mint	$25.00

BALLERINA BARBIE 1989 MINIATURE

A miniature Barbie doll in Ballerina fashion on a revolving stand with pink stage curtains at the back. Produced by Arco Toys Ltd., a Mattel Company.

©1989	Mint	$5.00

BATHING BEAUTY BARBIE 1959 MINIATURE #7478

A miniature Barbie doll in her original black and white bathing suit on a revolving stand with pink palm trees at the back. Produced by Arco Toys Ltd., a Mattel Company.

©1989	NRFB	$5.00

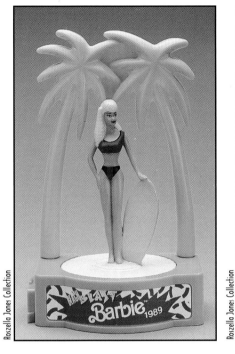

Clockwise from top left:

BEACH BLAST BARBIE 1989 MINIATURE　　　©1989　　　Mint　　　$5.00
A miniature Barbie doll in Beach Blast fashion on a revolving stand with lime green palm trees at the back. Produced by Arco Toys Ltd., a Mattel Company.

CALIFORNIA DREAM BARBIE 1988 MINIATURE
A miniature Barbie doll in California Dream fashion on a revolving stand with lime green fan on columns at the back. Produced by Arco Toys Ltd., a Mattel Company.
©1989　　　Mint　　　$5.00

COOL TIMES BARBIE 1989 MINIATURE
A miniature Barbie doll in Cool Times fashion on a revolving stand with a pink arch at the back. Produced by Arco Toys Ltd., a Mattel Company.
©1988　　　Mint　　　$5.00

DANCE CLUB BARBIE 1989 MINIATURE
A miniature Barbie doll in Dance Club fashion on a revolving stand with musical notes on a pink arch at the back. Produced by Arco Toys Ltd., a Mattel Company.
©1989　　　Mint　　　$5.00

Roszella Jones Collection

EVENING ENCHANTMENT BARBIE 1959 MINIATURE　　　©1989　　　Mint　　　$5.00
A miniature Barbie doll in Evening Enchantment fashion on a revolving stand with a pink fan at the top of columns at the back. Produced by Arco Toys Ltd., a Mattel Company.

Rozella Jones Collection

Clockwise from top left:

HAPPY HOLIDAYS BARBIE 1988 MINIATURE
A miniature Barbie doll in the 1988 Happy Holiday fashion on a revolving stand with a white arch at the back. Produced by Arco Toys Ltd., a Mattel Company.
©1988   Mint $5.00

SOLO IN THE SPOTLIGHT BARBIE 1959 MINIATURE   **#7478**
A miniature Barbie doll in Solo in the Spotlight fashion on a revolving stand with pink drapes and a mirror at the back. Produced by Arco Toys Ltd., a Mattel Company.
©1989   NRFB $5.00

Rozella Jones Collection

Rozella Jones Collection

Clockwise from top left:

SUPERSTAR BARBIE 1989 MINIATURE
A miniature Barbie doll in SuperStar fashion on a revolving stand with pink columns at the back. Produced by Arco Toys Ltd., a Mattel Company.

©1989   Mint $5.00

WEDDING PARTY BARBIE 1959 MINIATURE
A miniature Barbie doll in her 1959 wedding dress on a revolving stand with white columns at the back. Produced by Arco Toys Ltd., a Mattel Company.

©1989   Mint $5.00

BARBIE MOTOR BIKE   #4856
The motor bike comes with a working kickstand, safety helmet, books, and backpack. As the wheels spin, a realistic motor noise is made.

©1983   NRFP $35.00

Roszella Jones Collection

Roszella Jones Collection

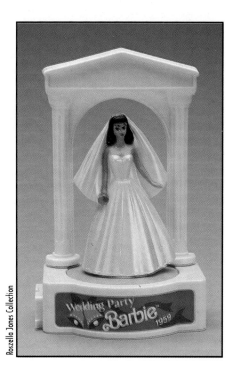

Roszella Jones Collection

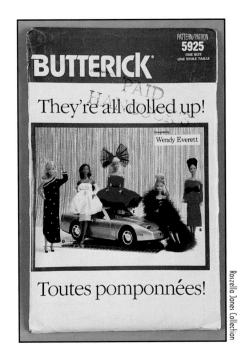

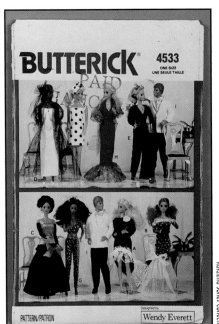

Clockwise from top left:

BUTTERICK PATTERN   **#5925**
Even though this is advertised as a pattern for an 11½" fashion doll, the dolls used as models are Barbie and her friends dolls. Designed by Wendy Everett and originally sold for $4.75.
©1987   Uncut $10.00

BUTTERICK PATTERN   **#6495**
Even though this is advertised as a pattern for an 11½" fashion doll, the dolls used as models are Barbie and her friends dolls. Designed by Wendy Everett and originally sold for $4.95.
©1988   Uncut $10.00

BUTTERICK PATTERN   **#3569**
Even though this is advertised as a pattern for an 11½" fashion doll, the dolls used as models are Barbie and her friends dolls. Designed by Wendy Everett and originally sold for $5.25.
©1989   Uncut $8.00

BUTTERICK PATTERN   **#4533**
Even though this is advertised as a pattern for an 11½" fashion doll, the dolls used as models are Barbie and her friends dolls. Designed by Wendy Everett and originally sold for $5.50.
©1989   Uncut $8.00

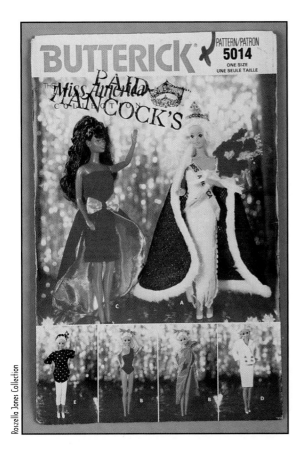

Roszella Jones Collection

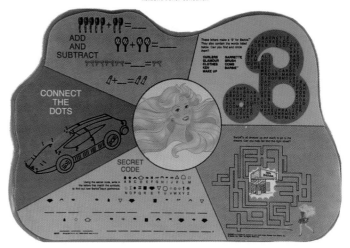

Clockwise from top left:

BUTTERICK PATTERN   #5014
Even though this is advertised as a pattern for an 11½" fashion doll, the dolls used as models are Barbie and her friends. It is the *Miss America Collection* and originally sold for $6.25.
©1990   Uncut $8.00

BARBIE PLACEMAT
This is a laminated placemat with artwork of Barbie doll in her bedroom in a purple party dress. The opposite side features four games.
©1988   Mint $3.00

Roszella Jones Collection

Roszella Jones Collection

Roszella Jones Collection

Clockwise from bottom left:

BARBIE PLACEMAT ©1988 Mint $3.00
This is a laminated placemat with artwork of Barbie doll roller skating with the beach in the background. The opposite side features four games.

PLATE HIGH FASHION BARBIE/THE 1959 BRIDE-TO-BE ©1990 Mint $43.00
This plate was produced by the Danbury Mint and features artwork of Bride-To-Be by Susie Morton.

PLATE HIGH FASHION BARBIE/THE 1960 SOLO IN THE SPOTLIGHT ©1990 Mint $43.00
This plate was produced by the Danbury Mint and features artwork of Solo In The Spotlight by Susie Morton.

Roszella Jones Collection

Clockwise from top right:

BARBIE FOR GIRLS PLAY BOX
This pink plastic box with handle features a photo of Barbie doll on the side.

©1991   Mint $3.00

BARBIE PLUSH DOG
This is a soft cuddly white dog with pink and purple accents featuring a heartshape adornment with the Barbie logo.

©1986   Mint $8.00

BARBIE FOR GIRLS PURSE
This is a vinyl, zippered purse. It is round with a heart containing a picture of Barbie doll on the outside. Pyramid produced a line of backpacks, bags, purses, and umbrellas with this same style.

©1991   Mint $5.00

100 PIECE JIGSAW PUZZLE ALL STAR BARBIE
By Western Publishing Co., Inc., this is a Golden 11½" x 15" 100-piece puzzle featuring All Star Barbie doll working out in the gym.

©1991   NRFB $5.00

Roszella Jones Collection

Roszella Jones Collection

Roszella Jones Collection

Rozella Jones Collection

Rozella Jones Collection

Rozella Jones Collection

Counterclockwise from top left:

COSTUME BALL BARBIE PUZZLE CARDS
These cards held eight pieces each to complete a 64-piece 8" x 10" puzzle featuring Costume Ball Barbie doll.
©1991   Mint $2.00

BARBIE FRAME-TRAY PUZZLE
Produced by Western Publishing Co., Inc., this is an 8¼" x 11" Frame-Tray puzzle featuring Barbie doll on a fashion runway in a fuchsia outfit.
©1984   NRFB $4.00

BARBIE FRAME-TRAY PUZZLE
Produced by Western Publishing Co., Inc., this is an 11¼" x 14¼" Frame-Tray puzzle featuring Barbie doll and friends in her bedroom playing records and eating popcorn.
©1985   NRFB $4.00

BARBIE FRAME-TRAY PUZZLE
Produced by Western Publishing Co., Inc., this is an 11¼" x 14¼" Frame-Tray puzzle featuring Barbie doll in a lavender party dress, crown, and pearls holding a bouquet of roses.
©1989   NRFB $3.00

Author's Collection

Clockwise from top right:

100 PIECE JIGSAW PUZZLE CRYSTAL BARBIE   #4609-43
By Western Publishing Co., Inc., this is a Golden 11½" x 15" 100-piece
puzzle featuring Crystal Barbie doll and Ken doll.
©1987   NRFB $8.00

100 PIECE JIGSAW PUZZLE HAPPY BIRTHDAY BARBIE   #4096
By Western Publishing Co., Inc., this is a Golden 11½" x 15" 100-piece
puzzle featuring Happy Birthday Party Barbie, Skipper and Ken dolls.
©1989   NRFB $6.00

100 PIECE JIGSAW PUZZLE ISLAND FUN BARBIE   #4096
By Western Publishing Co., Inc., this is a Golden 11½" x 15" 100-piece
puzzle featuring Island Fun Barbie, Skipper and Ken dolls in the tropics.
©1989   NRFB $6.00

Roszella Jones Collection

Roszella Jones Collection

Roszella Jones Collection

Roszella Jones Collection

Roszella Jones Collection

Roszella Jones Collection

Roszella Jones Collection

Clockwise from top left:

550 PIECE INTERLOCKING JIGSAW PUZZLE ▸ CASSE-TETE   ©1989   NRFB $22.00
Produced by American Publishing division of The Putnam Publishing Group, this is an 18" x 24" puzzle featuring Nostalgic Barbie 900 Series fashions.

550 PIECE INTERLOCKING JIGSAW PUZZLE ▸ CASSE-TETE   ©1989   NRFB $20.00
Produced by Western Publishing Company, this is a Golden 15½" x 18" puzzle featuring a group photo of Nostalgic Barbie 900 Series fashions modeled by the 1960s Barbie, Ken, and Midge dolls.

BARBIE AM RADIO   #3455   ©1980   Mint $10.00
This radio is in the shape of Barbie doll and fashion featured on its front.

BARBIE RADIO SYSTEM   ©1984   Mint $15.00
This pink and purple plastic, battery-operated radio has two speakers to complete the system. The radio is in the shape of the Barbie Portrait featured.

Roszella Jones Collection

Roszella Jones Collection

Roszella Jones Collection

Clockwise from top left:

2-PIECE BARBIE SCHOOL SET
This set contains a ruler and a purple vinyl, zippered pouch. It was produced by ADI, Inc.
©1983   NRFP $4.00

4-PIECE BARBIE SCHOOL SET
This set contains a ruler, eraser, pencil sharpener, and zippered vinyl pouch. It was produced by ADI, Inc. (See 2-piece school set for front cover of vinyl pouch.)
©1983   NRFP $5.00

COLORFORM® BARBIE LAZER BLAZERS
   STICKERS   #8602
This is a package of 3-D holographic stickers.
©1983   NRFP $2.00

BARBIE AND THE ROCKERS SOAP BOTTLE
This bottle is made by Ducair Bioessence, Inc. and originally sold for $4.00.
©1987   NRFB $6.00

Roszella Jones Collection

BARBIE LOGO ON PINK BANNER
This is vinyl banner that is 48" x 62" and was obtained from the now out-of-business Children's Palace.
©1988   Mint $35.00

BARBIE LANE RETAIL STORE SIGN
This is a hard plastic three-dimensional two-sided sign. The artwork is applied paperwork.
©1988   Mint $100.00

BARBIE PINK JUBILEE RETAIL STORE SIGN
This is a three-dimensional hard plastic and cardboard sign celebrating 30 magical years. It is 37" x 19" overall.
©1988   Mint $100.00

Roszella Jones Collection

Roszella Jones Collection

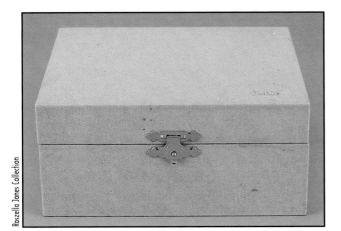

Roszella Jones Collection

Roszella Jones Collection

Roszella Jones Collection

Roszella Jones Collection

Counterclockwise from top right:

BARBIE 25TH ANNIVERSARY TEA SET

This tea set includes tray, four cups and saucers, sugar, creamer, and teapot. It is in a gray case with a silver metal clasp and hinges and has a Barbie logo and 25th anniversary imprinted in silver on the top.

©1990   Mint $78.00

BARBIE CHINA TEA SET 12-PIECE

This china tea set includes four cups and saucers, sugar, creamer, and teapot and lid. It was produced by Chilton Globe, Inc.

©1989   NRFB $20.00

BARBIE CHINA DINNER SET 16-PIECE

This china dinner set includes four plates, four cups and saucers, sugar, creamer, and teapot and lid. It was produced by Chilton Globe, Inc.

©1989   NRFB $28.00

Rozella Jones Collection

Rozella Jones Collection

Rozella Jones Collection

Clockwise from top left:

BARBIE TERRY BATH TOWEL SET
Produced by R.A. Briggs, this 2-piece set includes a terry bath towel and wash cloth.
©1990   Mint $15.00

BARBIE ROUND TIN
This pink 10 oz. tin came filled with bubble bath crystals. It features art-work of Barbie doll in a circular frame on the cover and was one in a line of gentle cosmetics for glamorous little girls.
©1988   Mint $4.00

BARBIE AND THE ROCKERS OUT OF THIS WORLD VIDEO
Produced by Hi-Tops Video, this fully-animated tape runs 25 minutes and features songs sung by Barbie and the Rockers dolls.
©1987   NRFP $15.00